Morphogenesis and Evolution

MORPHOGENESIS AND EVOLUTION

KEITH STEWART THOMSON

New York   Oxford
OXFORD UNIVERSITY PRESS
1988

Oxford University Press

Oxford   New York   Toronto
Delhi   Bombay   Calcutta   Madras   Karachi
Petaling Jaya   Sinapore   Hong Kong   Tokyo
Nairobi   Dar es Salaam   Cape Town
Melbourne   Auckland

and associated companies in
Berlin   Ibadan

Library of Congress Cataloging-in-Publication Data
Thomson, Keith Stewart.
Morphogenesis and evolution / Keith Stewart Thomson.

p. cm.
Bibliography: p.   Includes index.
ISBN 0-19-504912-8
1. Morphogenesis. 2. Developmental biology.
3. Evolution. I. Title.
QH491.T49 1988
574.1'7—dc19   87-24075   CIP

9 8 7 6 5 4 3 2 1

Printed in the United States of America
on acid-free paper

Preface

As I began to write this book while on sabbatical leave at the University Museum, Oxford, it was soon apparent to me that I had been working on it in one way or another ever since I was an undergraduate. It was as an undergraduate that I first read Waddington. Later, as a graduate student at Harvard, I was exposed to what was then the obligatory denigration of R.B. Goldschmidt and—perhaps typically—decided that if so many great scholars were spending time attacking him, he might have hit a raw nerve. Much later, at Yale, a group of graduate students and I decided to form a study group on alternative models of evolution to the population genetics bias of the "new synthetic theory" that we were learning and teaching. We called ourselves, rather self-consciously, the "heretical studies group." For two years, Stan Rachootin (now at Mount Holyoke College), Louise Roth (Duke), Amy McCune (Cornell), Scott Wing (Smithsonian Institution), Kevin Padian (U.C. Berkeley), Robert Nakamura (U.C. Davis), and I met and argued. One result of that was the paper by Rachootin and Thomson (1981) that forms the direct precursor of this book. To these colleagues and friends I owe an immense debt, but no blame should attach to them for this book's errors of omission and commission.

Tom Kemp in zoology and Tim Horder in anatomy at Oxford have helped me more than they may know, especially in being such generous hosts when I was at Oxford in 1984. I have ruthlessly sought out the advice of everyone I know to get assistance with this project. I am particularly grateful to two Englishmen in Wales, Robert Presley (Cardiff) and Richard Hinchliffe (Aberystwyth) for sharing their encyclopedic knowledge. As I have worked, I have realized not only how little developmental biology I know, but also how hard it is to master the complexities of the evolutionary side. Dolph Seilacher, Michelle Press, and Stan Rachootin generously read and criticized the manuscript, greatly to its improvement.

This is a book about a subject in transition—a subject that is still defining itself. The whole field of intercourse between developmental and evolutionary biology has a long and distinguished history, but it has only been recently and particularly in the light of hierarchical approaches to a range of evolutionary mechanisms, that a new, promising path through the maze of connections has seemed possible. People all over the world have now had their own "heretical studies groups," with the result that macroevolutionary approaches are respectable again and new roles for developmental processes in evolutionary mechanisms are being explored. The book has nothing to do with the old subject of "recapitulation," even though there is a lot that we still need to know about the causal connections between the patterns of ontogeny and phylogeny.

This book is intended primarily for evolutionary biologists, and most of the examples are taken from the vertebrates. Developmental biologists will find no new information on their subject, only a great abusing of their patience. They may, however, find interest in the uses to which developmental and especially morphogenetic information can be put. I have taken particular care to pepper the book with examples taken from the most current literature in developmental biology. The reader will find that the Bibliography is meant to be an introduction to the literature in the form of references to work relevant to the evolutionary theses of the book, rather than a review treatment with all the historical precedents carefully honored.

To discuss all of developmental biology in an evolutionary context would take a far greater knowledge of both development and evolution that I have. It would also take at least another ten years, for a lot of what is missing still has not been looked for, let alone found. In particular, I have not tried to incorporate in this small volume any discussion of the mechanisms of gene expression. I have deliberately concentrated on the later morphogenetic phases of development and their relationship to the processes of introduction of phenotypic variation at the level of the individual organism. This is not to say that control of gene expression is not a crucial element of developmental biology; rather, it is a different subject. Therefore, apologies for excluding these subjects would in part be false, because I hope to demonstrate that there are fundamental questions concerning evolutionary mechanisms that can only be asked and answered at the morphogenetic level. Eventually we will bring all the pieces together and make one whole theory and account of evolutionary mechanisms. But not yet.

A few passages in this book have been taken from work published elsewhere. I am grateful to Michelle Press (editor of *American Scientist*) for permission to quote from essays written for her journal. Even with this boost, for the last two years I have written this book during time stolen from my family in the evenings and weekends. Simply to offer my thanks seems trivial in comparison to what I owe to Linda, Jessica, and Rosey.

Philadelphia   K. S. T.
June 1988

Contents

Morphogenesis and Evolution

1
Introduction

All of science is fundamentally about cause. It is about explanations of the reasons things are the way they are and the mechanisms that produce them. It is now commonplace to observe that Charles Darwin brought evolution and all of organismal biology into line as a truly scientific subject by discussing evolutionary phenomena in terms of cause, and thus in the same testable, quantifiable frame of reference that applies to other science.

Darwin's theory of natural selection as a causal agency for evolutionary change was only the beginning of our problems, not the end. For more than a hundred years, we have sought to find all the layers and intersecting lines of causality that produce natural selection as well as to discover other mechanisms for change that are nonselective in nature—genetic drift or neutral mutations, for example. Natural selection is basically a mechanism that involves two components: the introduction of variants into a system and the subsequent sorting of these variants (Vrba and Eldredge, 1984) so that, over generations, there is a differential contribution of these variants to higher levels such as populations and species. Up to the present time, most attention of evolutionists has concentrated upon two aspects of the problem: the genetic basis of phenotypic variation and the dynamic properties of populations containing the individual variants. The present book is concerned with the mechanisms affecting the expression of variation among individual phenotypes. It has been a surprisingly neglected subject. The *New Synthetic theory* of evolution and its later modifications have largely been pursued as if the intrinsic mechanisms by which variation is caused among individual organismal phenotypes are less important to the processes of evolution than the extrinsic mechanisms of sorting. If only by default, variation introduced at the level of the individual phenotypes is commonly treated as if it were a simple mapping of variation at the genetic level, or at least were only a very simple function of that. It has seemed not only necessary but sufficient to study genetics in order to understand phenotypic variation. The reason for this must surely be that one tends to study those aspects of a science where a given concept or methodology opens the way; to paraphrase Medawar, science is the art of the possible. Even if one knows that there may be equally important aspects of the study that are being neglected, such knowledge is useless if one lacks the tools to engage with the problem. The situation is usually compounded by constraints of history (who has studied what) and style. In this case, quantitative and molecular genetics have been easier to study than mechanisms of development.

Phenotypes and therefore phenotypic variation have multiple causes. They are caused by genes and genetics. They are also caused by development; thus far, however, the processes and mechanisms of development—the "black box" standing between genetic variation on the one hand and phenotypic variation on the other—have not been considered a major part of evolutionary causality. If one looks in Mayr's landmark book (1963) for example, one finds that out of 673 pages, only some nine paragraphs are devoted to developmental questions, including the overworked question of the biogenetic law. The pivotal conference on and resulting book, *Genetics, Paleontology and Evolution* (Jepson, Mayr, and Simpson, 1949) contains no developmental biology at all. But phenotypic variation cannot be read out directly from genetic variation, and, as Kauffman (1983) has written, "only ignorance in general would have persuaded us that the response of any integrated complex dynamical system to random alterations in its parameters or structures would be fully isotropic." We have also to look at developmental processes in order to understand the generation of phenotypes. Genetic variation must be a foundation of all biological variation, whether seen at the organismal, deme, or species level. But it is time to examine in more detail what a view of developmental processes can add to the study of evolution, by opening up the black box and looking inside.

As discussed in the next chapter, the mechanisms of evolution can be seen to act at a series of different interactive levels from the genomic to the species level. Whereas Darwin concentrated on a mechanism that involved variation among individual organisms, variation must occur at several levels. Sorting processes—the equivalents of selection, drift, and so on—must also act at more than one level. The processes and results at each level will be discrete and different. We now discuss "species selection" quite comfortably, for example, as something quite different from individual selection. But the crucial stage in the whole suite of processes remains the generation of variation among individual organismal phenotypes.

Of course, it has long been obvious that genetic and phenotypic variation are at least partially uncoupled. Huge amounts of genetic variation are not expressed as phenotypic variation among individuals, as Kimura (1983) has shown. Some relatively small genetic variants, on the other hand, may produce quite major phenotypic change (homeotic mutants, for example). The mechanisms that underlie these asymmetries are not genetic but developmental. The properties of individual phenotypes are a product not merely of their raw genetic constitutions but also of the developmental processes by which the information of genetics is transformed into the flesh and blood of living functional organisms. These developmental processes have their own general rules and properties, and these rules and properties may be expected to add their own layer of causal influence in the whole evolutionary process.

The processes of development form a direct analogue of the processes of population biology. The properties of populations are modified, created, and controlled by processes acting at the population level, and much research has gone into discovering how these processes control the evolutionary mechanism. It is time now to add to our knowledge of evolutionary mechanisms by investigating the role of developmental processes in the creation of individual phenotypes and the causes of their variation.

This is not to say that work in other areas of evolutionary theory is so complete as to need no further attention, or that the roles of these areas will be superseded. Rather it is an approach to a more balanced theory of evolution in which causality at a whole range of different loci is investigated and the results brought into one synthetic whole. As will be outlined in the next chapter, the approach used here gives an explicitly hierarchical view of evolutionary mechanisms—one that shows us some of the main places where there are gaps to be filled and new opportunities to be seized.

The history of the study of evolution has been marked (as has that of other sciences) by a duality of approaches. One mode is to generalize, and compare, to look for the lawfulness and regularity of effect even though the mechanism is unknown. Here we see the search for the laws of morphology, so popular in the eighteenth and nineteenth centuries, and for the laws of development (to be discussed later) that dominated much of the nineteenth century. Here is where empirical comparative biology (embryology, morphology, functional anatomy, paleontology) is essential in terms of assembling the major and minor patterns that will serve to define important evolutionary problems. The second approach is to dig deeply into single mechanisms, pushing to the limit the particular methodologies and concepts on which such an approach always depends. Students using the latter mode are rarely interested in comparative biology and even tend to be highly skeptical of the explanatory power of other approaches. They usually seek to reduce explanation to the most atomistic level that current techniques can make available. The progress of science seems to consist of cycles in which these two modes alternate. Unfortunately, sometimes practitioners of the two approaches get confused and assume that they can do the whole thing by themselves. This was perhaps never more true than with the molecular chauvinism of the 1960s and 1970s. But it was certainly also true of those who hitched their wagon to the "biogenetic law" in its various versions in the last century, or to the quantitative genetics of the middle of the present century.

The study of developmental biology has been intertwined with that of evolutionary biology ever since Darwin, Muller, and von Baer. Both subjects deal with histories: that of the clade (phylogeny) and that of the individual phenotype (ontogeny). If detailed dissection of phylogenies can be thought to be useful in finding the lawfulness of evolutionary change, then what can the course of ontogeny tell us? The development of an organism is in a sense the supremely integrative process. It begins with the union of the germ cells and ends only with death. All the complex causes of the individual from molecular genetics to environmental effects, all the accumulated history of the clade to which the individual belongs, everything that is expressed in the phenotype and anything that is present but unexpressed, all these are brought together by the processes of development. It has always seemed therefore that, if one only had the cipher, the hidden language of evolution could be read out from the pages of ontogeny.

As it turned out, the limitations of the ontogeny–phylogeny parallelism were reached at about the same time that Mendel's genetics were rediscovered and, after a lull, attention of evolutionary biologists turned very strongly to the glimpses of mechanism afforded by study of genetic processes, particularly the quantitative genetics of populations. This lead to the formulation of the New Synthesis and its various modifications, including the new contributions of molecular biology, which holds the promise of providing a common language for all of biology.

A new phase of paying attention to new aspects of cause in evolutionary mechanisms has recently arisen in part because the New Synthesis appears to have reached a limit in its explanatory power. New attention is being paid to phenomena that New Synthesis has conspicuously failed to explain, especially those of so-called macroevolution. Macroevolution is a difficult term. Does it refer to taxonomic scale, morphological scale, or temporal scale? To many people it means a search for saltation and calls to mind the much-maligned work of Goldschmidt (1940) and "hopeful monsters." To others, it is the more readily acceptable search for phenomena and explanatory mechanisms at levels other than those of the individual and population. It is about phenomena such as long-term trends, and it always has asked nagging questions about the tempo and mode of evolution, and particularly whether all evolutionary phenomena are explainable in terms of gradual processes acting over very long periods of time in large or small populations.

A lot of the dissatisfaction that some students had come to feel with the New Synthetic or neo-Darwinian view of evolution was crystallized by the late Donn Rosen (1978) in a review of a book explicitly discussing natural selection in terms of population genetics. At about the same time, the whole of evolutionary theory came under challenge from a variety of sources, from anti-Darwinists such as Macbeth (1971) and from systematists who, like Rosen, found that current evolutionary concepts sometimes interfered with, rather than illuminated, discussion of relationships among organisms. The challenge came especially from Hennigian cladistic methodology (another case of a methodological breakthrough leading to the founding of a whole new subdiscipline, not always tolerant of followers of other, more traditional approaches and reinforced by the reaction of the latter). To its critics, the New Synthesis, and particularly the population-level approach, has tended to become a fascinating subject in itself but increasingly disconnected from anything except that which can be defined in its own terms.

As discussed in the next chapter, new hierarchical analyses of evolutionary mechanisms have had the salutory effect of pinpointing the gaps, and thus the opportunities, in our present view of evolutionary mechanisms. The aim of this book is to see where developmental mechanisms add significant new layers of causality in evolutionary mechanisms, to examine the mechanisms of developmental biology and see how they can explain properties of evolutionary processes—even to show that certain evolutionary phenomena can only be explained in such ways. It is therefore a strictly internalist approach that is discussed here. Obviously, evolutionary mechanisms are a mixture of internals (the causes of variation) and externals (sorting of variation). Both are important. But whether a new phenotypic variant survives is not just a matter of externals; it depends first on what the new phenotype is and does. Any new phenotype creates a new set of terms within which the externals operate. If it succeeds, it effectively defines its own selective regime out of the generality of selective regimes that applies to the larger group to which it belongs. Similarly, the course of evolutionary change over time is not entirely directed by external factors but is also heavily influenced by the biases and constraints of the internal mechanisms by which phenotypes are generated. It is with these matters that this book is concerned.

The book is arranged as follows. In the next chapter we will discuss more precisely how and where developmental mechanisms can affect evolutionary mechanisms. Then we will turn to discuss the nature of developmental mechanisms in three chapters dealing with the processes of pattern formation and morphogenesis. The seventh chapter aims to draw out some generalizations about the nature of morphogenetic systems and to show how these can affect evolutionary processes. Chapter 8 then attempts to lay out some aspects of the problems of evolutionary diversification that need to be explained. In the final chapter I summarize all this in terms of a particular set of evolutionary problems that are more fully explained by the addition of morphogenetic considerations and mention some of the many opportunities for future research.

2

Theory, Reduction, and Hierarchy

Implicit in the reasons given in Chapter 1 for development being ignored until recently as a potential causal factor in evolutionary theory is the general concept of reductionism. It is a strictly reductionistic approach either to believe that phenotypic variation is equivalent to genetic variation, or to act as though this were the case until disproven. Thus, to take but a single example, we find Stebbins (1974), who is avowedly a "strict reductionist," stating that "in the future all general theories about evolution will have to be based chiefly upon established facts of population and molecular genetics."

Reduction is, of course, a powerful tool, but it is one with which biologists have in general had difficulty, and which in recent years has come under strong attack and defense (see Williams, 1986). The basic reductionist statement with which we are all comfortable is ontological, namely that the processes underlying all living phenomena are reducible to the operation of mechanical causes: there is no irreducible vitalist essence. Reductionism in this sense is unexceptionable and universal in science. The more difficult sort of reductionism to deal with is theory reduction. A simple expression of this would be the statement that the laws of chemistry are all explicable in terms of the laws of physics, or the laws of biology in the laws of chemistry. Nagel (1961) shows that such theory reduction requires that, for example, the laws of chemistry must be deducible from the laws of physics and that the terms and concepts of both sets of laws be "connected" (see, for example, Newton-Smith, 1982; Beckner, 1974). Another way of putting it is that the laws of physics must be of wider scope than the laws of chemistry, which then constitute a series of special cases of the former, under particular boundary conditions.

Talking about theory reduction within the biological sciences, where general theories of broad scope are lacking (except the general theory of evolution that all organisms are related by descent), is somewhat pretentious. In the biological sciences we are forced to work more modestly with rules and probabilities rather than grand laws. The first step in unraveling schemes of causality is to find all the regularities, the patterns of effects, in the system. Then comes the more difficult task of discovering whether these are due to immanent properties of some particular set of causes or whether they are merely historically contingent. If they are the former, then they reflect rules or laws that can then be tested to see if they are reducible to the operation of more general laws (as a set of special cases) or whether they stand alone. Such is the case in trying to discover the role of developmental mechanisms in evolution.

It is this level of epistemological theory reduction that is most important to us, because it is only in terms of theories that the gaps between sets of knowledge (e.g., concerning genetic, phenotypic, and population characteristics in evolution) can be closed. A true reduction would be achieved if both the facts and the laws of phenotypic variation were totally explained as a special case of the facts and laws of genetics, under particular defined boundary conditions.

At once we can see a flaw in the "strict reductionism" of the statement just quoted. In fact, it is a statement that two sets of laws must be operating, the one not contained within the other. It is obviously the case that the facts and laws of population genetics are not totally and exclusively contained with the laws of molecular genetics. This is not to say that all populational characteristics lack a genetic basis, but rather to say that other rules and other factors than those that control molecular genetics must control population genetics (e.g., genetic drift, statistical probabilities). Molecular genetic mechanisms evidently will explain a lot, but only just so much. The question then becomes, what are the limits to any such explanation, and indeed what are the limits to the phenomena that are caused by the mechanisms of population genetics? In other words, if there are two sets of theories, molecular genetic and population genetic, that are useful to the study of evolution, are there more—in particular, the theories and laws of development? And how can we find out?

It is perhaps the most appealing notion of all science that explanations get simpler as science advances because science progresses by the formulation of ever more general theories. This notion persists despite the obvious corollary that the more general the theory, the more exacting must be the definition of the boundary conditions that produce the given phenomenon. This requirement makes science complex again. Even so, given enough information, it should be possible, say, to predict the freezing point of water from a knowledge of the physical chemistry of hydrogen and oxygen, or to predict the solubility in water of copper sulfate. Biological systems, however, are even more complex and, in particular, are interactive. Strict reduction can break down quite quickly. As an example we might take the case of epithelia. An epithelium consists of cells, but it does not consist only of cells. It also includes a basement membrane, which is a cell product. Now, it is not a general property of every group of cells to produce a basement membrane; they produce a basement membrane only in epithelia. Therefore, the basement membrane is a special feature of cells in epithelia (literally a special boundary condition!). But other properties of epithelia, such as the sheetlike ways in which they fold during development (and which can be so interestingly modeled; see Odell et al., 1981; Helfer and Helfer, 1983), are not properties of cells, but of sheets. They include properties held in common by sheets of paper, of cells, of molecules, thin films of liquids, and even gases. It is not a unique property of cells in epithelia to have sheetlike behavior. Therefore, an understanding of epithelia cannot be reduced simply to the study of their cellular constituents. The laws of epithelia are not simply a special case of the laws of cells under particular boundary conditions, and therefore in a basic epistemological sense such a reduction is impossible. But this is not to say that the explanation of epithelia can ignore the fact that they are formed of and by cells. Nor does it deny that epithelia are also composed of elementary particles, charges, and bits of space.

The nonreducibility of the properties and behavior of epithelia solely to the laws of cell biology is similar to the nonreducibility of population genetics to molecular genetics. In both cases we find that to unravel the situation we must start with a knowledge of the regularities, the rules, the laws of all the phenomena involved. It is here that the comparative approach in biology again comes to the forefront. If the cellular epithelium were the only example of a sheetlike phenomenon we knew, we might be tempted to try to find its explanation in the group behavior of cells and in turn to derive this from the behavior of individual cells. When we see the regularities among the behaviors of different sheetlike phenomena, we are forced to look outside the system. When we test the notion that something can be reduced to something else, we have first to define all the entities, phenomena, concepts, regularities, and laws that are included in the network of causality. But because one cannot know all these or the shape of their connections in advance, basically one must set up the analysis so as to be able to detect the existence of the unknowns.

This last is the greatest challenge in biology, with all its awesome complexity and interactiveness. Curiously, while it should lead workers to be cautious about causality, it seems to impel most biologists to an unquestioning reliance on a reductionist null approach. There are major problems with this. At its worst, a hypothesis such as "evolution can be explained by (reduced to) population genetics and molecular genetics" tends to lead to the naive (but arrogant) assumption that no form of investigation other than one couched in the terms and concepts of these subjects is necessary. If we work hard enough and wait long enough, everything will become explained. Sufficient information will become built up even to explain all higher level phenomena. Until then, students of higher level sciences (development or species-level phenomena) will be engaged only in housekeeping. This leads also to what might be termed the "two points make a straight-line graph" fallacy; one earnestly studies phenomena at two different levels (e.g., genetics and behavior) and then reduces the one to the other without regard to the intervening levels and intersecting phenomena.

Our task in evolutionary biology is to find all the cases where a true reduction applies (in the sense that the apparent laws of one system are only special cases of laws acting at a more general level) and all those cases where independent phenomena operate.

The resistance of biological systems to naive reductionist approaches is a product of their interactive complexity. The properties of a biological phenomenon are produced by sets of interrelated causes. In trying to sort out such complexities, biologists tend naturally to head for hierarchical analyses. Such analyses are logically dictated by the time-linearity of complexity in biological phenomena: the linearity of simple to complex in individual development (fertilized egg to adult), and the linearity of simple to complex transitions in lineages of phenotypes in evolution. Hierarchy is, of course, explicit in all classifications, and classifications are the foundation of comparative biology. A whole range of hierarchical schemes can be devised to deal with, for example, ecology, behavior, or structure. These hierarchies can have their own structural properties—principally those of nested sets and part–whole (Salthe, 1986; Eldredge and Salthe, 1985; see also Ghiselin, 1987). A hierarchical analysis gives us separable levels of complexity and thus a scheme to find where different concepts and laws operate and how they might be related. A hierarchical analysis ought therefore to show us the potential role of development in the causality of evolutionary change. Whether or not developmental mechanisms actually contribute to evolution will then have to be solved through an examination of the facts and particularly the rules of development itself. But first we have to show that there is a place where they could operate in evolution.

HIERARCHIES IN EVOLUTIONARY ANALYSES

In recent years several authors (including Eldredge, 1982; Vrba and Eldredge, 1984; Eldredge and Salthe, 1985; Salthe, 1986; Ghiselin, 1987), building upon an older foundation (see Grene, 1987), have applied concepts of hierarchical analysis to the study of evolution. Although many aspects of the subject and its application to evolutionary theory are still controversial, it is possible to isolate some simple general principles that are of great value in the process of discovering the potential role of developmental processes in evolutionary mechanisms.

We may start by defining a working hierarchy, a hypothesis of a set of phenomenological levels at which processes occur and at which, at least potentially, separate sets of rules are in operation. As Vrba and Eldredge (1984) discuss, evolutionary mechanisms always involve two fundamental processes: the introduction of variation and the sorting of variation (selective or nonselective). We are well accustomed to thinking in such terms when we consider variation at the level of individual organisms. Variation is introduced into the system in the form of phenotypic variation among individual organisms. Sorting of this variation then occurs so that some particular subset of the individuals that potentially might contribute to succeeding generations actually does so. This is introduction and sorting at a single operational level. In a hierarchical view of evolution we seek to identify a series of different levels at which comparable processes of introduction of variation and sorting occur. In addition, at each correctly defined level in a hierarchy there will be a set of properties that are unique to that level. Individual organisms make up demes and populations. However, a true deme or population is more than a chance assortment of individuals. It is defined by breeding systems, patterns of gene flow, balances of polymorphisms, degrees of heterozygosity, and so on. Thus at the population level there are properties that are more than simply those due to the sum of the constituent parts (individual organisms). This is crucially important because it means that the rules of demes and populations are different from the rules that govern the biology of individual organisms.

Just as there is variation among individual organisms, so variation occurs between or among populations, each of which has different characteristics. There is also sorting among populations so that some survive and others do not. Therefore, in this set of statements we have described two phenomenological levels—the organismal level and the population level—at both of which comparable processes operate. Vrba and Eldredge (1984) proposed that a complete theory of evolution would involve processes of introduction of variation and sorting acting in a hierarchy of different levels—what they term a "genealogical hierarchy." In it we can define four and possibly five levels: at the lowest level are the genomic constituents (see, however, Ghiselin, 1987); then individual organisms, demes, species, and, conjecturally, higher taxonomic levels beyond species.

Each of these levels can potentially contribute to an evolutionary mechanism, through introduction and sorting of variation. To take introduction of variation: at the gene level it occurs through mutation, recombination, and any other factors that cause changes in the genetic codings of development. At the level of the individual organism, variation occurs in terms of phenotypic characters. Among demes it exists, for example, in differences in gene frequencies and breeding system characteristics, as just mentioned. Among species there may be differences in a host of features from specific mate recognition signals to morphological or physiological characteristics. Each of these types of variation, expressed at a particular level, will be caused in one way or another by processes acting at lower levels. For example, phenotypic variation is obviously genetically based. But it is not only genetically based. There are other factors involved—environmental factors, for example—that only become expressed at the focal level of the organism. Similarly, demic variation is caused by mechanisms acting both at the levels of the genotype and the individual organism, and in addition by processes acting only at the deme level (e.g., breeding systems). Furthermore, no matter what is the sum of causes that produces the variation introduced at any level, at each level the variation is always manifest in a way that is particular to that level. Genetic variation is expressed only at the gene level, phenotypic variation only at the individual level, demic variation only at the deme level, even though each will reflect processes acting at preceding levels.

Hierarchical analysis will only work if we have chosen the correct phenomena and assigned them to appropriate levels. All the phenomena must be strictly comparable and be capable of discussion in terms of statements of the same logical type. In the case of hierarchical analysis of evolutionary mechanisms this means, as noted above, that at each level processes of introduction and sorting of variation must occur. Similarly, if what seem to be two levels actually do not have separate properties, then we cannot assign them different stations in the hierarchy. For example, one might want to argue about whether kin groups could occupy a place between the levels of individual organism and deme. We will not get into such a discussion here, but it can only be resolved by examination of the special rules and properties of kin groups as compared with individual organisms and demes.

Therefore, the second criterion for judging the appropriateness of a hierarchy is whether all the levels are definable in the same terms. One would not insert the category "organ" between "gene" and "individual organism" in a genealogical hierarchy, although we could devise a different hierarchy to examine different processes for which a morphological hierarchy such as molecule–cell–organ–organism might be appropriate. Similarly, one could establish an ecological hierarchy (e.g., organism–population–community) to examine yet another set of biological questions. The hierarchy gene–organism–population–species–higher taxon is a "nested set" of levels at which processes of introduction of variation and sorting of variation occur. It is not a hierarchy of part–whole composition. Each level is defined by the level(s) below, but, for example, individual organisms are not composed of genes.

There has been a great deal of debate within the evolutionary literature concerning "levels of selection," that is to say, the hierarchical levels at which selection can act. Logically selection can only act upon certain categories of phenomena. For example, individual organisms are the paradigm level upon which selection acts. An individual organism either contributes to the next generation or it does not, and whether it does so depends on some function of what it is. At this point we run into a semantic confusion concerning the use of the word "individual." Most students of the problem believe that in a general sense of logic and rhetoric, we need to distinguish those entities that function as "individuals" and those that are "classes." An example often given is that the element gold is a class. All atoms with a particular defining characteristic (the atomic number 79) are atoms of gold. Each atom, however, is an individual. Progressing hierarchically, however, the gold is also a member of the class "elements." A long-standing debate in evolutionary theory (starting, perhaps, with Lewontin, 1970; and continued by Ghiselin, 1974; Hull, 1980; Sober, 1986) concerns which categories of phenomena count as "individuals" rather than simply having class membership (and indeed about the value of trying to make the distinction). For the purposes of trying to lay out a hierarchical view of the evolutionary process, the most pressing question has been whether species count as individuals or not. Various criteria exist for qualification as an individual in this sense. Individuals have discrete beginnings and ends (births and deaths) and are spatiotemporally organized. Complex individuals have a particular pattern of organization and it is organization, rather than the sum of the parts, that causes the properties allowing the system to qualify at an individual at a particular level—the level at which these properties are expressed. Thus, if a deme is to act as a focal level in our hierarchy, it must be defined not simply by the sum of the properties of all the individual organisms constituting it, but by the relationships and interactions between them that hold the deme together, that make it work, make it even recognizable, as a population. Such properties are "emergent."

Hierarchical analysis is usually thought to require that each focal level has the formal status of individual. Probably individual organisms and demes are readily accepted as individuals. There has, however, been much debate over the question of whether species are. Traditionally they have been treated as classes, because of the somewhat artificial way in which they are usually recognized in practice (through commonality of their parts), and because of the difficulty in seeing them as spatiotemporally bounded. The traditional view has been that species grade insensibly into one another over space and time and that distinctions among them are taxonomists' artifacts. But even granted this, species may function as individuals if they have their own characteristic properties of organization or assembly that produce features not definable simply as the sum of their parts. Again, one such feature would be their breeding system. The consensus view now seems to be that species can act as individuals and therefore that the species level is an appropriate component of our hierarchical analysis. It is a fascinating question, whether there are transspecific focal levels as well. If there are evolutionary processes that operate at the species level, are there any that operate beyond? Obviously there are transspecific patterns, by which we recognize higher taxonomic categories. But in order to qualify as levels for evolutionary mechanisms to operate, these (or other more subtly defined categories) must be the site of both introduction of variation as well as sorting. The current view would be that transspecific patterns are merely the result of historical sorting.

A hierarchical model of evolutionary mechanisms can thus be proposed in which, following especially the work of Vrba and Eldredge (1984), four (and possibly a fifth) levels of causality are recognized on the basis of the following features in common. They are all levels at which processes of introduction of variation and sorting can occur. They are specifically recognizable as "individuals" rather than "classes." They are spatiotemporally bounded, self-replicating, and with definable beginnings and ends. They are complex, with special emergent properties arising from their intrinsic "rules of assembly" or organization. The special properties thus arising are expressed only at that particular level—which we will call the focal level—and thus are quite different from properties that arise simply as the sum of the constituent parts, for these are expressed at the lower levels.

CHARACTERISTICS OF HIERARCHIES

The most important feature of a hierarchical analysis is the most obvious: setting out the hierarchy forces one to isolate and then examine all the components of a complex system. The hierarchy that has just been outlined is probably overly simple, but it at once demonstrates that most accounts of evolutionary mechanisms are rather incomplete. For example, the standard neo-Darwinian/ New Synthesis account of evolution is inadequate in that it deals only with the introduction of variation and the sorting of variation at a small subset of the levels at which mechanisms must be in operation (Figure 1). Setting out the hierarchy forces us to examine whether important processes might be occurring at the other levels as well. Until such processes are defined and studied, the hierarchy remains essentially a hypothesis, but this kind of analysis ensures that our hypothesis is a broad one. Indeed it is by its nature an expandable

Focal level	Process	Phenomenon		
		From lower	At focal level	
Organism	Introduction	1	2	
	Sorting	?	3	NST
Population	Introduction	1, 2, 3	4	NST
	Sorting	?	5	NST
Species		1, 2, 3, 4, 5	6	
		E	7	SS

Figure 1 Operation of different phenomena at three levels of a hierarchy: individual organisms, deme/population level, and the species. 1 = genetic factors; 2 = development; 3 = selection and sorting of individuals; 4 = population/deme assembly processes; 5 = population/deme sorting processes; 6 = species level assembly processes; 7 = species level sorting processes. NST = domain of the New Synthetic Theory; E = domain of Effect Hypothesis; SS = domain of species selection theory.

one. We can look for new levels in between the existing ones, or at either end of the hierarchy.

The second important advantage of such hierarchical analysis is also rather obvious. It demonstrates the interactiveness of the complex system that forms that causation of evolutionary mechanisms. Not only can there be, in fact there must be, "simultaneous operation of processes among individuals at several levels" (Vrba and Eldredge, 1984). In addition, the results of processes at each level directly influence processes at other levels; they contribute to the causation at other levels.

Introduction of variation and sorting of variation can occur in three principal modes. First and most obviously, they occur at each focal level and must occur at all focal levels. But equally, processes occurring at one level have a causative influence at other levels, either above ("upward causation") or below ("downward causation"): the terms are due to Campbell (1974). Furthermore, upward or downward causation may be due to either the process of introduction of variation, or its sorting, or both.

Upward causation is perhaps the easiest to visualize. From the preceding discussion it will already be obvious that the effects of all phenomena at lower levels in the hierarchy will be carried forward in the same generation willy-nilly to the upper levels. An example of upward causation from the genomic level would be the proliferation of "selfish DNA," which is an introduction of variation at that level. This selfish DNA will be taken along for the ride to higher levels. For example, massive gene duplication in amphibians and lungfish is correlated with large cell size, slower cell cycles, and slower metabolic rates. Similarly, sorting at the genome level, in the form of selection with respect to different genetic environments, will affect the total genotype and thereafter affect the introduction of variation at the level of the individual organism to the extent that genome factors are involved in the complex of processes that produce phenotypes (i.e., through development). Upward causation,

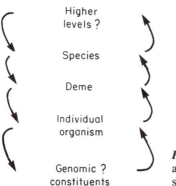

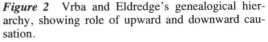

Figure 2 Vrba and Eldredge's genealogical hierarchy, showing role of upward and downward causation.

whether the result of processes involving introduction of variation or of sorting at a given focal level, is always manifest at higher levels as effects on the introduction of variation phase. Downward causation is perhaps less familiar. But obviously any process of sorting or introduction of variation at a given focal level must potentially produce upward causation in the same generation and then cause downward in the next generation by reference of the results of the operation of the whole hierarchical mechanism back to the genomic level (Figure 2).

THE POTENTIAL ROLE OF DEVELOPMENT IN EVOLUTION

From the preceding, the potential importance of developmental processes in contributing to the total array of causes of evolution change becomes obvious. Development literally is the process of organization and assembly of the individual organism. Development is therefore the immediate cause of introduction of variation at the level of the individual organism. The processes of development are processes acting at the level of the individual organism, combining genetic and environmental information in the production of each individual. The common rules of development (those that are not simply special cases of the rules of lower level phenomena) are the organizing rules for this focal level: they cause those focal-level characteristics that qualify it for inclusion in the hierarchy. We can postulate that the unique properties of each individual in part represent upward causation from the genomic level plus the unique environment and circumstances in which each developmental scenario is acted out. The common properties of all individuals within a given taxon, expressed both in terms of common genetic information and common environmental experiences, represent an element of downward causation from higher hierarchical levels.

The aim in dealing with complex mechanism is to discover those unique properties and processes that cannot be understood except at the appropriate focal level. Central to understanding the role of development in evolutionary mechanisms must be the study of the emergent and epigenetic properties of developing systems and their unique role in the processes by which variation is introduced among individual phenotypes. A quite dramatic example of the role of morphogenetic processes in distinction from raw genetic information is provided by Provasoli and Pintner (1980; also Tatewaki, Provasoli, and Pintner, 1983), who showed that the familiar "sea lettuce" *Ulva lactuca* and the related green alga *Monostroma oxyspermum* did not produce their typical thallus when grown in a bacteria-free culture; instead they grew in masses of filaments. The polymorphism turns out to be environmentally rather than directly genetically mediated.

If developmental processes are in fact important in evolutionary mechanisms, the sort of hierarchical analysis briefly summarized here places the subject in a context in which it can be scientifically examined, and on the same terms as other components of the complex set of mechanisms that cause evolutionary change. At the same time, the hierarchy also gives a hypothesis to falsify. If development is not important, we should be able to focus investigations to find that out.

The basic questions to be asked are as follows:

1. Do developmental processes significantly filter or modify that component of variation among individual phenotypes that is directly ascribable to genetic differences alone? In other words, is there an asymmetry between genetic and phenotypic variation?
2. If this is the case (as all contemporary experience would seem to confirm), then in what ways can developmental processes affect phenotypic variation and what are the mechanisms by which an asymmetry between genetic and phenotypic variations is caused?
3. What particular phenomena would this sort of causality uniquely explain?
4. What would the consequences of such causality be for our views of evolutionary processes in general?

Having mapped out a general field of enquiry, we need now to turn to look in some detail at the nature of developmental processes. In this it will become obvious that "development" is also a highly complex system that can be analyzed hierarchically, and we will have to keep the two hierarchies, evolutionary and developmental, clearly distinguished before we can see how they combine.

3

Development:
Pattern and Process

In a hierarchical system, the manifestations of pattern at one level tend to become the components of process at the next, an alternation that is repeated again and again. The processes of molecular genetics produce, under the special circumstances applying at the genomic level, a pattern of gene activity that is carried forward to the organism level. Here new developmental processes based on these gene-level patterns build new organism-level patterns, and these then become the raw material for deme-level processes acting at the next level. This duality of pattern and process operates in any interactive system and was addressed particularly by Gregory Bateson (1979). It requires that we discuss the phenomena of process and pattern separately, as well as discovering their interdependence especially when, as in biological systems, there is additional complexity present in the form of feedback of causations among levels.

Implicit in any hierarchical analysis is the assumption that process always involves a lawful set of mechanisms. In all biological systems this lawfulness will be derived from two sources: from the immanent properties of the systems themselves, and from even more general laws applying across all biological systems. For example, the salivary glands and lungs of vertebrates are both constructed in part according to a strict set of developmental rules applying to mechanics of epithelia; these are examples of developmental constraints (Chapter 7). But the size and shape of the lungs also follow more general physical rules such as the gas diffusion laws or volume–surface area relationships that determine how big a lung is required for an animal of a given size. These are structure–function constraints. Biological systems also derive a major set of consistencies from the historical connectedness, through relation by descent, of the organisms concerned. These consistencies are often called "phyletic constraints" (Chapter 7).

It is a basic approach in biology to use the analysis of pattern to approach an understanding of process, often first in terms of deriving the "rules" from study of consistency and regularity. This is where the great power of the comparative–analytical method lies. Comparisons of the patterns produced by a mechanism give us the first stage in getting to grips with the process itself, and in fact a considerable amount of progress can be made toward defining the elements of the process in this way when techniques for direct analysis are still unavailable. The classic example of this is, of course, the study of genetics.

18

Starting with Mendel and then all the great advances of the early twentieth century, quantitative analysis of the patterns of organismal and population-level genetics revealed a great deal about the probable nature of the mechanisms of heredity itself. Quite significant progress in this area was possible without knowledge of the actual molecular biology of genes.

The history of biology, as is the case with most sciences, has had a rhythm that has depended on particular technical approaches in the direct analysis of mechanism. In developmental biology, by the earlier part of this century, experimental embryology seemed to be yielding important results, but these approaches eventually reached a technical limit and by the 1950s "embryology" was no longer considered a forefront science. Techniques that have been spun off from the surge of interest in molecular biology have now reopened and revitalized the study of development, principally at the subcellular and cellular rather than the morphogenetic end of the developmental spectrum. Life naturally tending to maximum contrariness, it is the morphogenetic level of development with which we need most to be concerned as evolutionists. But there is enough new work available to begin to open up the old questions relating the process and pattern of development to the mechanisms of evolution.

EVOLUTIONARY AND DEVELOPMENTAL PATTERNS

Attempts to find regularity and lawfulness in developmental data occupy a very special place in the history of modern biology. One automatically recalls the nineteenth-century attempts to find laws of parallelism, from the "threefold parallelism" of Agassiz to the explicit attempts to link ontogeny and phylogeny, starting with Oken and others and most permanently associated with the names of Meckel, Serres, von Baer, Muller, and Haeckel. If nothing else, this great body of work has had tremendous heuristic power in a science in which some kind of connection between development and evolution seems obvious, but at the same time remains as elusive as a butterfly. The "biogenetic law" of Haeckel, in particular, seemed at one time to be a powerful biological generalization. Remnants of the biogenetic theory still permeate biology: for example, one often finds reference to evolution as a process involving "terminal addition," a straightforward Haekelian concept.

Attempts to find a scientific relationship between ontogeny and phylogeny are, however, fundamentally limited. Despite the best efforts of evolutionists, it is impossible to escape the fact that ontogeny is a concept of "form changes" over time: ontogeny is a pattern rather than a process (discussion in de Beer, 1958). Therefore, it can only be compared with other concepts of pattern such as form changes over evolutionary time (phylogeny), if at all. Ontogeny is not a process or a mechanism of development; therefore, it cannot be translated into concepts of mechanisms of evolution. And, of course, it was strictly in such formal terms that Meckel, Serres, and other early students of ontogeny and phylogeny worked. They were formulating laws of parallelism between the patterns of anatomy seen in different frames of space and time.

We do not need to review here the history of study of ontogeny and phylogeny; it was brilliantly surveyed by Russell in *Form and Function* (1916) and reconsidered in explicitly evolutionary terms by de Beer (1958), whose work was in turn revised and extended by Gould (1977). We should, however, note the different nature of the two main approaches (see Patterson, 1983). Von Baer's conclusions concerning the regularities and lawfulness of ontogeny are summed up in his four laws: (1)"That the general characters of the big group to which the embryo belongs appear in development earlier than the special characters"; (2) the less general structural relations are formed after the more general, and so on until the most special appear"; (3) "the embryo of any given form, instead of passing through the state of other definite forms, on the contrary separates them"; (4) "fundamentally the embryo of a higher animal form never resembles the adult of another animal form, but only its embryo" (von Baer, 1828, translated by Huxley).

On the other hand, the early law, usually termed the Meckel–Serres law after its joint proponents, is more explicitly recapitulatory (Meckel, 1821):

The development of the individual organism obeys the same laws as the development of the whole animal series; that is to say, the higher the animal, in its gradual evolution, essentially passes through the permanent organic stages which lie below it; a circumstance which allows us to assume a close analogy between the differences which exist between the diverse stages of development, and between each of the animal classes.

Haeckel's later expansion of the whole subject (1866) is summarized in the biogenetic law:

Ontogeny, or the development of the organism individual, being a series of form-changes which each individual organism traverses during the whole time of its individual existence, is immediately conditioned by phylogeny, or the development of the organic stock (phylon) to which it belongs.

Or,

Ontogeny is the short and rapid recapitulation of phylogeny, conditioned by the physiological functions of heredity (reproduction) and adaptation (nutrition). The organic individual . . . repeats during the rapid and short course of its individual development the most important of the form-changes which its ancestors traversed during the long and slow course of their palaeontological evolution according to the laws of heredity and adaptation.

Patterson (1983) shows that the different developmental laws involve quite different consequences for the reconstruction of phylogeny.

Haeckel very carefully qualified his phrasing of the biogenetic law. He was a superb embryologist and knew that the available data did not support a one-to-one mapping of ontogeny and phylogeny. "The complete and accurate repetition of phyletic by biontic development is obliterated and abbreviated by secondary contraction, as ontogeny strikes out for itself an ever straighter course; accordingly, the repetition is the more complete the longer the series of young stages successively passed through." And "the complete and accurate repetition of phyletic by biontic development is falsified and altered by secondary adaptation, in that the bion during its individual development adapts itself to new conditions." Russell (1916) points out that these two provisos are taken directly from Muller.

Thus Haeckel understood that von Baer's first and second laws are often violated. A simple example can be given from vertebrate development. A crucial feature of vertebrates compared with, say amphioxus, is the formation of a whole prechordal region of the head, the forebrain and associated structures in particular. In amphioxus the notochord reaches right to the tip of the snout. In all vertebrates it reaches only so far forward as approximately the line of the hypophysis. In vertebrate development, the forebrain is induced in the neural tube by the underlying prechordal plate, a section of mesoderm that is located at the front of the chordamesoderm and that at gastrulation invaginates immediately in front of it physically and before it temporally (Chapter 5). The prechordal plate, having performed its various inductive roles, then disperses and goes on to form part of the mesenchyme that will give rise to the skeleton of the anterior braincase. The prechordal plate differentiates before the presumptive notochord, but the notochord is phylogenetically the more primitive character. If von Baer's first and second laws were universally true, then the prechordal plate, which is evolutionarily more derived (assuming that the condition in amphioxus is primitive), would have to differentiate after the notochord. That it does not do so has to do with the mechanisms of early vertebrate development. Because gastrulation puts presumptive regions into their correct places, and because the regions of chordamesoderm that organizes the neural tube and other structures must all invaginate at gastrulation, those invaginating first must be differentiated in advance of later invaginating structures. Phylogeny, the order in which various innovations happen to have been acquired, is subservient to the demands of the mechanism itself. Von Baer's third and fourth laws, however, are fundamentally correct and the timing of the prechordal plate and notochord in development just given is evidence of that fact. (One could, of course, put all this in reverse if amphioxus were secondarily derived from a vertebrate through appropriate modification of the relative position of the chordamesoderm and neural plate at gastrulation.)

Haeckel's biogenetic law and his concept of evolutionary change through terminal addition and condensation have long since been discarded. We can chart their demise both in terms of their own scientific inadequacy demonstrated by the forceful arguments of von Baer, Sedgwick, or Garstang, and through the rise in fashion of a direct approach to the causal mechanisms of evolutionary change—population genetics. Sedgwick's observations perhaps serve as an epitaph: "A species is distinct and distinguishable from its allies from the very earliest stages all through development, although these embryonic differences do not necessarily implicate the same organs as do the adult differences" (1894).

There is however, one place where the study of the pattern of form changes may offer direct evidence concerning the sequence of evolutionary changes through simple mechanisms affecting pattern itself. This is the phenomenon of heterochrony (de Beer, 1958). The study of heterochrony grows out of a parallel tradition—a line connecting the works of d'Arcy Thompson (1917), Huxley (1932), and de Beer. Heterochronic evolutionary changes occur through alteration of the processes controlling the patterns of size and shape change and the timing of maturation. Gould (1977) epitomized the relationships among these in a clock model. Heterochrony can be used to explain the evolution of the allometric shape transformations that d'Arcy Thompson demonstrated could account for the apparently drastic differences among the phenotypes of many animals. A nice case is comparison of the shapes of skulls among adult and juvenile apes and humans. Gould (1977, figure 61; cf. Stark and Kummer, 1962) shows that adult human skull shape is in fact readily seen as a neotenic version of the fetal chimpanzee skull shape.

Heterochrony is a powerful explanatory mechanism of the cause of a class of evolutionary changes. The mechanisms underlying heterochrony essentially involve control of the rates at which a small range of established developmental programs are expressed. Because it does not involve whole-scale qualitative reorganization of the mechanisms of development, one can even use the pattern of ontogeny to chart hypotheses of the pattern of phylogeny in cases where the changes have occurred relatively recently. However, despite the claims that are made on behalf of heterochrony as a general factor in evolutionary change, the number of cases in which heterochronic mechanisms are the sole or principal cause of significant innovation, as opposed to within-group diversification, are probably rather small. Heterochrony is a valuable concept because it links development and evolution in fairly understandable ways. It is a mechanism of change that seems to work; yet it works only within existing pathways. It does not wholly explain what de Beer (1958) carefully distinguishes as change through deviation in ontogenetic pathways or what I refer to later as developmental reprogramming. Once one comes to analyze a complex sequence of evolutionary changes, even one in which size and shape transformations appear only to involve some relatively simple allometric relationships, one finds that major deviations are involved and these are not explained by heterochronic resetting of the clock, but by reprogramming of the underlying processes. An example would be Radinsky's study of ontogeny and phylogeny in the horse skull (1984). The change in architecture of the horse skull over time, which seems superficially to be largely the allometric consequence of size increase, in fact involves several significant qualitative reorganizations of skull morphogenesis.

Nineteenth-century embryologists were looking for very general laws. Our aims in attempting to link processes of development with the processes of evolution are perhaps less grandiose. Developmental mechanisms contribute only to a portion of the complex of processes that cause evolutionary change. The key to attempting to understand the mechanisms and unraveling their connectedness will be first to understand the patterns and processes of development and then to relate these to observable evolutionary phenomena.

THE RULES OF DEVELOPMENTAL PROCESSES

In the following chapters we will attempt to discover those consistencies, regularities, rules, or laws of the processes of development that will allow us to explain or predict the ways in which developmental mechanisms operate within the hierarchy of causes to contribute to the nature of evolutionary change. Certain features are obvious in advance.

1. Developmental processes are very conservative. They change progressively—that is, without major inconsistencies—because all known developmental processes have evolved by modification from existing processes and mechanisms. This is the explanation of the pattern of consistency that is observable in the evolution of phenotypes and the appearance of congruity between ontogeny and phylogeny. It is the explanation of homology.
2. This consistency fundamentally stems from inherent historical and material limitations or constraints operating on the possible ways in which both the genetic system and internal or external environments can themselves change.
3. Patterns of change in the phenotype will be partially uncoupled from the pattern of potential change coded in the genotype to the extent that the developmental mechanism itself has buffering properties. Any complex system, especially one with apparently as much redundancy as the genetic mechanism of higher organisms, must be highly buffered and this will be manifest in two ways. There will be simple buffering—suppression of potentially major disturbances of the system. But buffered systems also show threshold effects. Therefore, we should expect that any change in a developmental mechanism might produce either directly proportionate effects or disproportionately large and small phenotypic effects. (Of course, this also depends on what one terms "large" or "small.") As study of heterochrony shows, a small number of allele substitutions affecting genes that control rates of processes potentially will have a greater effect on the phenotype than the same number of substitutions in "structural genes." But equally, one substitution at just the right time and place could potentially cause a threshold effect radically changing a whole developmental pathway.
4. The development of the embryo is essentially autonomous. While there may be various strong environmental influences, the embryo controls its own development. A major portion of the history of developmental biology has been concerned with the question of whether these processes are largely deterministic or regulative. That is, whether development is a process that is controlled by a set of intrinsic factors and, once started, simply unfolds according to a proscribed and rigid course, or whether the very processes of development produce a capacity for self-correction. Obviously, the extent to which deterministic or regulative processes operate in development will greatly affect the potential for phenotypic change—for example, the capacity for buffering just mentioned.
5. Development will be controlled in part by the rules applying to its components. There are rules of materials, of structure and proportion, rules of epithelia and rules of mesenchyme, rules of cell movement and cell division, and so on.
6. Development also involves major interactions among the different component systems within the embryo, each also controlled by rules. These are, particularly, the rules of cell-to-cell contact, relationship of cells to sub-

strates, and of inductive interactions. There are rules of gradients, of waves, and of timing. These produce higher level rules of pattern control, timing, relative growth, and allometry.

We need to understand developmental phenomena sufficiently well to discover the modes and tempi of introduction of phenotypic variation into the nexus of causal evolutionary mechanisms. This means that we must understand a huge range of phenomena starting with the foundations of genetics. In these chapters I will concentrate on morphogenesis, the later stages of the development of phenotypes. This is partly because the subject is otherwise too large to encompass in one small book, partly because it is the field I know better, and partly because it seems to be in morphogenesis that many of the most interesting mechanisms in the cause of phenotypic changes actually occur. Before we discuss morphogenesis, however, we must first set the stage by considering some aspects of the earliest stages of development.

4
Early Pattern Formation

The processes of development form a continuum that begins with gametogenesis and ends only with the death of the individual organism. It is therefore artificial to try to define separate phases and stages of these processes, just as it is artificial to try to separate the structural history of the embryo into a series of discrete forms through time with discrete and definable properties (let alone trying to match such artifacts to putative phylogenetic stages). But at the same time, the sequence of mechanisms of development is hierarchically organized. The major early event is the transfer of control over development itself from the purely maternal factors inherited within the egg and particularly in the egg cytoplasm, to the switching on of the zygotic genome and transfer to zygotic control of cell function, interaction and differentiation, morphogenesis and cytodifferentiation. This transfer does not occur at a single instant, nor is it easy to generalize about it even with a single group of organisms.

Other landmarks are harder to find, especially ones that can be compared consistently over a range of different organisms. However, one can roughly divide the processes of development, for the purpose of organizing a discussion at least, into two main phases: early and late pattern formation phases. Early pattern formation can be defined as that part of the developmental sequence in which all the major mechanisms that control the shaping of the embryo, both its morphogenesis and cytodifferentiation, are set into place. In a vertebrate, early pattern formation would be everything from gametogenesis up to and including gastrulation, by which time all the essential elements of tissue interaction that will cause the morphogenesis of the embryo have not only been regionally defined and correctly positioned, but have started to function. Late pattern formation comprises the stages of morphogenesis and cytodifferentiation. As we will see, morphogenesis itself can be divided into two stages, early and late: roughly speaking, in the earlier part, morphogenetic pattern-controlling mechanisms are set in place, and in later stages their results are expressed (Figure 3).

Obviously these are still very arbitrary definitions. Early pattern formation is that part of development in which, referring back to von Baer's first and second laws, the general features of the major group of organisms to which the taxon in question belong, have been set in place. In vertebrates, early pattern formation is thus that stage in which the major defining characters such

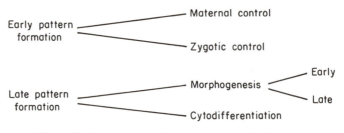

Figure 3   Terminology for the stages of development.

as the notochord are first discernible in rudiment. Cohen (1979) calls this early phase the "phyletic stage" of development. This is probably a useful distinction, but it is essential to emphasize that the structures of the embryo that one can see at this early phase are to be viewed more as fundamental elements to the process of development than as rudiments of adult organs. What is important about the chordamesoderm in early development is not what it is going to become in the adult, but that it has a major early role in inducing other structures of the embryo. This is a theme to which we shall return again, but is necessary to emphasize here that we need to discuss embryonic phenomena in terms of processes, not structural patterns.

SPECIFICATION, COMMITMENT, DIFFERENTIATION, AND DETERMINATION

In order to proceed with the discussion it is necessary to define four terms that will be important for the whole of this book. First we need carefully and accurately to distinguish among the terms specification, commitment, and differentiation. These terms describe both states and the processes by which these states are acquired. Together they encompass the process of determination of cells. A cell lineage is "specified" to have a certain fate very early in the normal development of any animal. By using fate maps and other techniques we can employ a sort of hindsight to show that cells in a particular region of the embryo will normally follow a certain fate; they are specified for that fate. They may not make it. They may still be diverted in an emergency to other ends, especially in response to experimental insult. If tissues at this stage are excised from the embryo and cultured they may not yet have the capacity to form those particular structures without further instructive signals, but their normal fate is nonetheless specified. Next, at some point early or late in development, the history of those cells includes a number of processes and events that make them more and more committed to that particular fate. They are becoming committed, normally irreversibly, by a process essentially of restriction of the number of options open. The process may occur at a single step or progressively. For example, a region of cells already irreversibly fated to become mesoderm will in the course of development become further regionalized and specialized into myogenic, nephrogenic, sclerogenic mesoderm, and so on.

"Commitment" involves the control of unique gene expressions and is a moving position between specification and the last stage, which is differentiation. "Differentiation" is the final stage of irreversible commitment to a particular cell type—chondroblast or osteoblast, for example. Finally, "determination" is a term referring to the process by which levels of commitment—that is, progressive gene expression—are caused. A determinant is a specific factor causing a particular type or phase of gene expression.

REGULATIVE AND MOSAIC DEVELOPMENT

The whole business of trying to set out formal categories is essential to the making of the essential comparisons from which we can try to draw general conclusions concerning the regularities, rules, and laws of development. But it tends often, rather than freeing us for discussion, to make us a prisoner of the terms that we have created. In perhaps no case in biology is this more apparent than with two terms that are routinely used to describe the nature of early pattern formation. These terms are "mosaic-determinate" and "regulative-indeterminate." They have always been useful metaphors for description and discussion of major differences in the process and pattern of development among different major groups of animals. But they have also become concepts that confine as well as facilitate discussion.

"Mosaic" and "determinate" are terms that have applied to a mode of development in which the egg cleaves and cell lineages are produced according to a strictly prescribed program from which deviations are not possible. From even before the first cleavage of the fertilized egg, in fact from the formation of the oocyte, a set of strictly determining conditions is spatially fixed. These involve physical factors that dictate an irreversible fate for all resultant cells. Examples that have traditionally been given are the development of nematodes or molluscs.

In "regulative" or "indeterminate" development, as the name suggests, the fate of cell lineages, while it may be quite strictly specified and predictable in normal development from topographic position within the developing embryo, is not actually fixed until much later stages. The definitive instructions prescribing the fate of given cells and their descendent clones are given late rather than early in the sequence of development. Experimentally induced errors (such as deletion of a group of cells) can be accommodated, adjusted for, or "regulated" by the embryo. Instead of the fate of every part of the embryo being irreversibly determined in advance, they are actually only acquired as development proceeds. Indeterminate embryos are not only self-regulating but, in every sense, self-assembling. Control over the processes of development is created progressively rather than being pre-determined in the oocyte and fertilized egg. The classic examples of regulative-indeterminate development are seen in vertebrates and particularly mammals, in which the embryo actually forms from only a smallish part of the total volume of cells produced by cleavage of the egg.

"Mosaic-determinate" and "regulative-indeterminate" are terms that apply to the early pattern formation stage of development. Beyond that point, once the zygotic genome is completely activated, the fates of all cell lineages in all animals (plants do things very differently), with certain very specialized exceptions such as the interstitial cells in cnidarians, are settled. However, up to this point, there are apparently great differences to be investigated and these differences have important sequelae for considerations of evolutionary theory. They illuminate the whole question of the control of early pattern formation. Therefore, although, as we shall see, the distinction is out of date, it is worth pursuing for a moment.

TESTS OF DETERMINATE OR REGULATIVE BEHAVIOR IN EARLY PATTERN FORMATION

As the preceding discussion suggests, some fairly simple criteria can be used to compare the processes of development in any given taxon in terms of the determinate–versus–indeterminate polarity. Examining some of these tests emphasizes the heuristic value of the distinction.

Regionalization of the Egg

Conceptually the simplest but technically one of the most difficult ways in which to discover the way in which the embryo develops is through studying the basis of regionalization in the egg. This can be accomplished indirectly through the effects of strictly localized surgical deletion, radiational damage, or biochemical blocking techniques on the cytoplasm of the fertilized or unfertilized egg. Perhaps the most dramatic example of mosaicism is provided by experiments on the fertilized egg of squid (Arnold, 1968; Marthy, 1975). In cephalopods (in distinction to all other molluscs) the egg cleaves to form a blastodisk laid out over a yolk mass. This blastodisk develops by fairly straightforward regional differentiation into the ground plan of the adult. A precise fate map can be drawn showing the spatial mapping of adult structures onto the blastodisk surface. Experiments with local radiational (UV) damage and surgical deletions show that the same fate map exists on the cortical surface. A particular regional disruption can be tolerated by the embryo, which continues to develop, but the affected region will be missing. It is possible, for instance, to produce embryos lacking the left or right eye by making the correct deletion in the egg. Unfortunately, the technical difficulties of this sort of work are tremendous, and Graham and Waering (1984) cast doubt on the results.

A limited example of a similar result can be obtained in amphibian embryos. In anuran amphibia, but not urodeles, the future germ cells of the adult can be shown to be mapped as a particular region of the vegetal cytoplasm of the fertilized egg. UV radiational damage in this region will produce an otherwise normal, but sterile, adult lacking germ cells and can be rescued by implantation of vegetal pole cytoplasm (Zust and Dixon, 1977). It is interesting that this does not occur in urodeles, in which development of the germ cells is therefore regulative.

Separation of Blastomeres

Here the results can be quite conflicting. In a fully mosaic embryo, if the first two blastomeres were separated and cultured apart they should each form the appropriate half-embryo. The closest we get to this is the classical case of Reverbi and Minganti's work on the eight-cell tunicate and Wilson's work on molluscs where early blastomeres, if isolated, produced appropriate tissue types when cultured. In regulative forms, whole embryos can result from separated first blastomeres, although in many taxa a significant proportion of the resultant embryos may be undersized. For example, work on amphibians shows that the first two blastomeres can develop normally if isolated. But if the first cleavage plane is altered experimentally so that the two blastomeres do not contain the normal proportions of material from the animal and vegetal halves of the egg (Chapter 5), they fail to develop normally. Thus normal development requires a correct distribution of materials contained within the egg cortex. In amphibians, isolated blastomeres from the four-cell and eight-cell stage never develop normally. Again, this must be because they lack the correct mixtures of cytoplasmic derivatives from other parts of the embryo. These cytoplasmic materials are not necessary strictly to specify particular structures but rather to set up the correct pattern of conditions for development of the (highly regulative) embryo (see discussion below and next chapter).

Culture of Half-embryos from beyond the Two-Cell Stage

It has been known since the work of Spemann that if one divides an amphibian embryo into two halves at the four-cell or eight-cell stage, the results once again depend on the plane of separation. When the half-embryo contains a combination of both animal and vegetal half-cells, development is normal. This will occur if the separation is in the line of the original first cleavage plane. Dorsal or ventral halves alone cannot develop normally. Recent work by Kageura and Yamana (1983) has shown that the conditions of culture also have a significant effect on the results. If lateral half-embryos (which traditionally have been expected to develop normally) are cultured in a special medium, half-embryos are formed in about 20% of cases. Despite thoughts that this might provoke concerning a possible inherent mosaicism of the amphibian embryo (see, for example, Cooke and Webber, 1983), this seems to occur because in this particular culture medium, the open wound of separation does not heal and the experimental animal does not become completely enclosed within an epithelium. Therefore, the experimental half-embryo seems to behave, as it were, as if the other side were still there. This immediately recalls Roux's famous experiments on the two-cell blastula where, if one cell is merely killed, but not separated, the other cell develops into a half-embryo, again acting as though the other half were present. In normal culture in Kageura and Yamana's experiments, where the wound healed over, then the two halves formed one whole each.

Much information concerning the nature of control of early developmental stages has been gained from studies of partial embryos, containing less than half a normal embryo, of defective embryos where only a small portion is removed, and fusions of isolated blastomeres and of partial embryos. In each case we find embryos differ in the extent to which the phases of development have been irretrievably determined in the oocyte or fertilized egg stage or are capable of regulation at later stages. There are perhaps two classic extremes, the mosaic role of the polar lobe in molluscan development and the regulative power of the dorsal lip of the blastopore in amphibians demonstrated in Spemann and Mangold's double-embryo experiment. But most developmental phenomena turn out not to be so easily classified.

BREAKING DOWN THE DISTINCTION

As I have remarked elsewhere (Thomson, 1983), pairs of apparently polar opposite terms like mosaic–regulative, determinate–indeterminate, form–function, or pattern–process usually turn out not to be true opposites in the sense of forming the ends of a single linear spectrum, but are complementary. They might be thought of as representing axes in a multidimensional space, with the phenomenon to be described occupying a point or space (usually moving during its own history), with respect to these axes.

The developmental processes of most organisms turn out never to be fully determinate or regulative. Instead, as many authors (including Huxley and de Beer, 1934) have discussed, most organisms at certain times and in various parts show both mosaic-determinate and regulative-indeterminate behavior. Most organisms, and perhaps all, show phases of development that could be described as regulative, at least in later phases of development. An example would be the nematode *Caenorhabditis,* which Sulston et al. (1983) show has the capacity to regulate experimentally induced loss of certain blastomeres. Another case is the pulmonate gastropod *Lymnaea palustris,* which is an equally rather than unequally cleaving mollusc. In reviewing evidence for regulative development in this member of a phylum in which mosaicism has been thought to be the rule, Arnolds et al. (1983) conclude "that the gastropod embryo follows a rigid mosaic, but that the embryo also has regulative capacities with respect to the form of the head, and it appears likely that this will hold for other adult structures also" (see also Morrill, Blair and Larsen, 1973; van Dam and Verdonk, 1982).

On the other side, there is often some sort of determinate behavior in regulative development. For example, Johnson et al. (1984) emphasize the mosaic nature of the early mammalian embryo. We will return to this discussion after we have further considered the nature and basis of these different behaviors. But in the meantime it is worth nothing that these distinctions between apparent rigidity and flexibility must have significant evolutionary implications, if the control of development is linked to the pattern of introduction of phenotypic variation (see Chapter 7).

THE BASES OF DETERMINATION AND REGULATION

The value of discussion of the mosaic-determinate versus regulative-indeterminate problem is that it forces us to focus not only on the course but also on the causes of determination in early developmental stages. One of the most fascinating trends in modern developmental biology is a reversion to a distinctly "deterministic" approach to the subject as opposed to the "epigenetic" approach that has dominated for perhaps 50 years or more. It is part of a general trend toward deterministic (and reductionist) views of biology that seem to have swept the subject since the arrival of modern molecular methods of studying the genetic code. These methods opened up a whole new range of material explanations of fundamental biological processes: processes that it had hitherto not been possible to study so directly. With concentration on a reductionist genetic base, deterministic thinking has been particularly prevalent in the sociobehavioral side of biology and in development. Reductionist-deterministic thinking is exemplified, for example, in the essay on developmental strategies by Cooke and Webber (1983) and in the determined (so to speak) efforts to show that mammalian embryos after all do have some mosaic properties. It may even be reflected in the efforts of Jacobson (1982; then immediately retracted, 1984) to demonstrate that the Spemann double-embryo experiment depended on extensive migration and reorganization of previously committed neurectodermal cells, rather than new induction by the organizer transplant: Smith and Slack (1983) have clearly shown this idea to be incorrect.

The positive result of this renewed focus on the deterministic processes on which reductionist approaches to development can be based is, of course, that a great deal of new effort has been directed to the important question of discovering the mechanisms by which factors present in the egg control the course of early pattern formation. Early pattern formation, being the process by which the early embryo becomes spatially ordered and a diversity of genetic expressions in different tissue types begins to appear, requires the existence of spatial order, in fact of spatial information.

Determinism in the very early embryo must result from strict spatial ordering (literally a mosaic) of discrete factors, expressed first in the oocyte, reinforced and reordered at fertilization, and then precisely distributed to the cells produced by cleavage. The planes of cleavage must be then strictly oriented with respect to this structuring of the fertilized egg. The materials distributed to the resultant blastomeres must be either qualitatively or quantitatively different, cell by cell. Their function must be to control the behavior of the cell in terms of the planes of further cleavages and the further subdivision and distribution of cytoplasmic factors to the daughter cells. This is not to say that the cytoplasmic factors actually control and cause cell commitment directly. But they must at least set in motion processes that cause such commitment. Cytoplasmic determinants may have a causal effect on the physical behavior of the cells during early morphogenesis (rates of division, patterns of motility, adhesion, response to extracellular matrices, for example). The final result is the activation of those regions of the zygotic genome appropriate to morphogenesis and eventual cytodifferentiation. At the same time, of course, all cells must inherit from the fertilized egg whatever gene products are necessary for their normal metabolic functions until the zygotic genome takes these over also.

The oocyte and fertilized egg are enormously complex structures, and the range of this complexity is still becoming understood. For example, the oocyte in most taxa contains huge populations of RNAs necessary for supporting protein synthesis in the early embryo. If these had a discrete spatial ordering (perhaps mediated by the cytoskeleton), they would form a basis for determinate development. The cytoskeleton itself obviously has some of the properties required, especially because it forms a consistent spatial array within the cell. But it is becoming clear that no one class of molecules or structure can be the exclusive or even the principal agent. Materials like yolk granules may be very simply distributed in the egg according to relative density gradients. Nonetheless, the spatial ordering of small to medium-sized molecules on the cytoplasmic skeleton of the oocyte offers a mechanism for very precise deterministic behavior. Perhaps the most complete work in the whole area has been done on ascidian development, where the famous yellow crescent in the egg has attracted a great deal of attention. Jeffery (1985) has shown that, contrary to expectation (e.g., Jeffery and Meier, 1983; Cooke and Webber, 1983), there is no preferential distribution of mRNAs in the yellow crescent. There are, however, a dozen or so proteins, some connected with the cytoskeleton, that can be shown to be unique to the yellow crescent domain.

In regulative development, layers of control processes are built up along with the layers of structural complexity because they not only produce this complexity but are produced by it. It is a spiral of complexity feeding into complexity, depending at all times on "whole-organism" properties as well as the separate regional specification of subparts. However, just as the embryo of any given taxon may show both determinate and regulative properties at different times of development, so all regulation must depend on a prior set of reference points in order to get started. While the response may be to a gradient, the gradient itself can only be set up via fixed points. The whole process forms a series of cycles. Fixed points provide the basis for regulative development, which thereby creates a new set of conditions that serve as fixed points for another round of regulative development, and so on. Therefore, even the initial spatial ordering of the cytoplasmic determinants (in the oocyte) must reflect some whole-organism (i.e., whole-cell) properties and must involve the capacity for some self-regulation, particularly if differential distributions of "morphogen" molecules are involved. Once the determinants are in place, however, their ordering and their control over development may be fixed until it has become diluted later in development (see Arnolds, Van der Biggelow, and Verdonk, 1983).

MATERNAL VERSUS ZYGOTIC CONTROL OVER EARLY PATTERN FORMATION

The major part of the control of very early pattern formation in all animal embryos seems to be under the control of the maternal rather than the zygotic genome. The exact time of switching on of the zygotic genome so that it can take over control of embryonic development differs from taxon to taxon. In many cases it still is unknown. It would be useful to have information on the time and conditions at which the zygotic genome starts to become activated, in a broad range of taxa. Transcription itself does not always mean that the genes in question are actually functioning in the cell; there are many further layers of control that may affect the process. But at least the detection in the polysomes of mRNAs from the zygotic genome would be the first set of data to deal with. (The 5S and 18S ribosomal RNAs have been studied in considerable detail, but they offer a different view of the state of embryonic gene activity—the control of the embryonic metabolism by the maternal versus zygotic genome.)

The difficulties are multiplied by the fact that the oocyte may come equipped with its own vast battery of mRNAs, and the activity and replication of these in early stages is difficult to sort out from the creation of new populations of zygotic origin. However, control is possible through use of hybrids, where the paternal genome provides a marker that is only expressed when the zygotic rather than maternal gonome is involved.

Once again the best data are available for sea urchins and amphibians. Gurdon's work on the ribosomal RNAs in *Xenopus laevis* shows that new populations of RNA come rapidly into existence at about the midblastula stage; see also Korn (1982). Galau et al. (1976), in a classic paper on the sea urchin, demonstrated that transcriptional activity begins very early after fertilization, but that translational activity and actual control by the zygotic genome do not occur until much later. Remarkably, in sea urchins, the zygote genome gets very fully switched on early, but some genes that are transcribed never get out of the nuclear envelope. This is in contrast to most vertebrates, in which the genome is more slowly activated and more immediately translated (Meuler and Malacinski, 1985; Sargent, Jamrich, and David, 1986). In mammals, where the egg is relatively very small, first zygotic gene expression occurs at the two-cell stage (West and Green, 1983; Bolton, Oades, and Johnson, 1984)

Another view of the control over early pattern formation and the timing of transfer from maternal to zygotic control comes from the data that have accumulated in support of the "differential gene expression theory." The original question concerned the manner in which the zygotic genome controlled development from morphogenesis to cytodifferentiation. There were two possibilities: either the genome was differentially switched on (and often off again) during development but the genomic constitution of every cell remained the same, or different cell lineages inherited materially different genetic constitutions, perhaps through selective deletion of particular regions of the genome. If the latter were true, then development would be mechanically irreversible. If the former were true, then the irreversibility of cytodifferentiation or earlier stages would all be a matter of the regulation of gene activity—what genes were selectively made active, rather than what genes were present.

The critical tests were provided by the work of Briggs and King (1952), King and Briggs (1956), Curtis (1962), and Gurdon (1977) using amphibians. Spemann (1938) had started the field off by experiments in which he enucleated a fertilized egg and transplanted into it the nucleus from a 16-cell-stage embryo. This resulted in normal development. The recent work extended the line of experiment in an effort to find out what sort of determination of nuclear functions had occurred at which stages of development. The results showed that a nucleus from any stages up to the beginning of gastrulation could initiate normal development, unless it was taken from that portion of the blastula that corresponded to the presumptive dorsal lip of the blastopore of the future gastrula. Thereafter, once any cell has passed over the dorsal lip of the blastopore, its potential range of functions have become restricted. If, for example, a nucleus from the presumptive endoderm of the early-gastrula stage were implanted into an enucleated zygote, normally the result was a rudiment with only generalized endodermal properties. This work then confirmed the differential gene expression hypothesis and indirectly confirmed that during early amphibian development the zygotic nucleus lies dormant, as it were, until it starts to become activated at the mid- to late-blastula stage. Once gene activation begins, cell commitment is normally irreversible, although dedifferentiation of adult nuclei is possible under certain culture conditions.

THE CONTROL OF REGIONAL ORGANIZATION

The determinate or regulative processes of early pattern formation get the right cells into the right places and in the right juxtapositions. This produces an array of cells with the correct range of sizes, shapes, surface properties, and behavior including motility and patterns of division, plus a precisely localized array of extracellular matrices, all of which are needed in order for morphogenesis to proceed. That is, it lays down the foundation for new processes of further pattern formation growing out of the existing arrangements and processes. Put simply, early pattern formation provides the information for the initiation of further gene expression, specifically that of the zygotic genome.

As we have discussed above, the information upon which early pattern formation is based comes from a variety of sources. It comes more or less passively from the distribution to daughter cells and clones of cytoplasmic determinants from the oocyte. It also comes from localized signaling at cell-to-cell contact and interaction and from differential cell behavior. Control over pattern formation results from the establishment of, and response to, whole-organism properties such as fields and gradients, and from the setting up of global or more localized patterns and reference points.

In much of this, the subregions of the oocyte or fertilized egg and the cells of the early embryo act as if they had "knowledge" of their positions in both time and space. For example, Horstadius has shown (1973) that the cleavage patterns of sea urchins follow a strict internal clock that seems to run according to an absolute rather than a relative time mechanism. The case of position "sense" is even more interesting, however, because no matter whether we are dealing with the most strictly mosaic-determinate or the most fully regulative-indeterminate development, the "right" cells and cell lineages will normally always be in the "right" place.

Wolpert (1969, and subsequent papers) has articulated the concept of "positional information" according to which cells are capable of two important functions: (1) They must register their position in the embryo according to a set of signaling systems and then "remember" this information. (2) Subsequent cell behavior is set in motion by response to this signaling system and the cells then differentiate either autonomously or through addition of interactively shared information. The information systems involved in early pattern formation include those just mentioned as determining factors. The polar coordinate model of French (1976; Bryant and Bryant, 1981) is based primarily on regeneration studies, particularly in insect imaginal disks and limb regeneration in amphibians, and offers a different and more specific model for the control of positional information. There are many difficulties with the concept of positional information and the model of polar coordinates, although the phenoma they deal with are real. Each certainly seems to apply in particular situations, but whether they are general descriptions of developmental processes in all organisms remains to be seen.

Positional information must represent a response to pattern-generating and controlling mechanisms operating in the embryo, from egg to late morphogenetic stages. Turing (1952) was one of the first to attempt a theoretical analysis of how combinations of simple phenomena like gradients and waves could impose a precise yet dynamic ordering in developmental systems, building up complexity out of an apparently disordered and naive embryo. This work has been greatly extended in recent years (e.g., see Goodwin and Cohen, 1969; Meinhardt, 1982). This system of pattern generation and control itself changes throughout development; at each stage it creates the raw materials out of which the next level of pattern control will operate (see Wolpert, 1983; and Chapter 7).

Concepts like positional information very naturally tend to cause one to focus upon specific determinant causes—special morphogen molecules, for example. However, the crucial information may just as easily be contained within the spatial ordering of components whose primary function has nothing to do with spatial determination per se. A signaling system consists of two elements: signaling and responding. Nothing is a signal until something else has responded to it. Therefore, the information may as likely be in the receiving as in the sending system. This is classically the case in the endocrine system where evolution of the system largely involves changes in the receptors. A similar case seem to apply in the all-important processes of induction.

Induction is a process in which the control of regional differentiation, instead of depending on mechanisms of clonal history (mosaic determination) or long-distance, whole-organism properties (positional information), can be viewed as arising from "purely local and progressively arising circumstances" (Horder, 1983). Most of the interesting work on inductive interactions in development has been done in vertebrates, which are marked by two major inductive phenomena: the primary organizer, which is the dominant component of early pattern formation in all chordates (see especially the discussion of Medawar, 1954); and epithelio-mesenchymal interactions, which are responsible for a great majority of morphogenetic processes in vertebrate organogenesis (Chapter 6).

Induction is a process of cell-to-cell interaction. It can occur either through short-range diffusion of signal substances across intracellular spaces, or through direct cell contact. The signal and response can only occur within a given time "window"; if induction does not occur in this interval it will be lost. A major effort has been devoted to the discovery of specific inductor signal substances transmitted within the embryo, summaries are found in Saxen and Toivonen (1962) and more recently in Toivonen (1979) and Nakamura and Toivonen (1978). Frankly, the results are disappointing for those intent on discovering specifically determining inductor molecules. Even the most instructive of inductive interactions can be triggered by whole ranges of foreign signals. Using artificial inducers, Tiedmann (1967) had success in selectively inducing from amphibian blastula forebrain, hindbrain, or mesoderm and was able to mimic a double-gradient model of gastrulation. A large number of workers has combined to show that the role of the primary inducer in amphibian gastrulation can apparently be mimicked by extracts from essentially everything from boiled liver to "adder's fork and blind-worm's sting, lizard's eye and owlet's wing." This leaves the investigator feeling very much like Macbeth on the blasted heath, whose fate was accurately predicted, but who was unable to control it.

The response of cells to induction is, of course, new layers of gene activation and RNA transcription. If the responding tissue is treated with substances to block RNA synthesis, the inductive response does not occur.

The signal and response of inductive interaction are local phenomena. Where, then, do the signaling and responding cells get their prior instructions from? They must arise either from successions of localized inductive-type signal–response interactions, or they must be "set" at some point by response to global conditions (gradients, or reference to fixed points, for example). It is almost possible to imagine a scheme of fully regulative development in which all control of regionalization occurs through induction and in which the pattern of controls simply assembles geometrically with increase in complexity, never requiring any global frame of reference. However, in real organisms there must at critical times (or even continuously) be an overlay of global instructions that keeps the ramifying network of inductive interactions in the correct relationships by specific reference to whole-organism properties.

THE SIGNIFICANCE OF DETERMINING PROCESSES

Evidently some embryos are set up to function perfectly with a system in which a majority of pattern formation is prescribed in the egg stages by maternally derived cytoplasmic determinants. Others, like the amniote vertebrates, function with a system of almost complete regulation. But perhaps no taxon has a development that is exclusively of one or the other type. Cohen (1979) suggests that the presence of a maternal blueprint in the egg protects the early stages of development and the crucial stages of pattern formation from error due to heterozygosity and the presumed mutational load of the zygotic genome. He calls this the maternal buffering effect. A maternal mosaic forces early development into a pattern that will not be swayed by any irregularities stemming from the activities of the zygotic genome. This, of course, means that any inheritable change in the early pattern formation stage, generation to generation, must be caused within the maternal side of inheritance only. However, this insulation of the embryo from the vicissitudes of the zygotic genome merely pushes the problem one stage back. It seems inescapable that the mosaic control system must be set in place, at least in part, by a self-regulating process that can accommodate "errors" in control arising within the maternal genome or the maternal environment. It simply cannot be left vulnerable to any chance variation because the mosaic system, once set in place, allows no further readjustments. At the same time, once it is in place, it is immune to further buffetings, internal or external, until the revolution produced by the switching on of the zygotic genome, which produces a new and potentially overriding set of controls. And even this event is controlled by the mosaic system.

In terms of any evolutionary potential, therefore, one should expect that a determinate system will always adhere rather closely to a particular ground plan that can be changed only with considerable difficulty, while regulation based on the zygotic genome might perhaps be more free to work a series of minor late pattern variations on this theme.

Regulative development is ideal for very simple or very complex systems. For a certain range of complexity mosaicism may be better, but there are limits to the amount of complexity that can be fully encoded in the oocyte and fertilized egg. Once that limit is reached, further information and control has to be added in layers, the coding for which has to be developed out of what already existed in preceding stages, which means regulation. For example, we have noted the traces of regulative behavior in gastropod mollusc development (Arnolds, van der Biggelow, and Verdonk, 1983). In a more fully regulative system like that of vertebrates, there is perhaps no upper limit to the amount of complexity that can be coded for. Regulative development has the advantage of being buffered against internal or external disturbance by the active epigenetic processes, which react against and adjust to, rather than being insulated against, potential disturbance. But this then also gives a potentially more labile system. There is considerably greater potential for nonlethal disruption, which is then a potential for evolutionary change to occur at almost any stage.

5

Example: Early Pattern
Formation in Amphibia

The Amphibia has been one of the most important animal groups for the study
of developmental biology, and a huge descriptive and experimental literature
has accumulated over the years. While sea urchins, molluscs, and nematodes,
and more recently, *Drosophila,* have each become an important vehicle for the
study of different aspects of development, the Amphibia and chordates in general have been especially important as the vehicle for the study of inductive
regulative mechanisms. The early development of all chordates is marked by
two revolutions in the control of early pattern formation: dorsalization at the
blastula stage and gastrulation—primary induction caused by the "organizer"—which was studied in great detail in Amphibia by Spemann and his co-
workers and continues to be the subject of intense scrutiny. The early phases
of development in Amphibia exemplify rather well some of the problems in
discovering the causal processes in development, whether in the egg, at fertilization, in the blastula, or in gastrulation itself. The short discussion of early
development in Amphibia that follows is meant to exemplify rather than catalogue these questions.

ORGANIZATION OF THE EGG

The oocyte in amphibians is radially symmetrical. A first axis of symmetry is
established between a more heavily pigmented animal hemisphere and a less
pigmented vegetal hemisphere. Before fertilization the egg is covered with a
transparent vitelline membrane. When fertilization occurs, the vitelline membrane lifts from the surface of the egg and the egg is then free to rotate inside
it so that the animal hemisphere lies uppermost and the vegetal hemisphere is
lowermost. This rotation is apparently a response to gravity, which means that
the vegetal hemisphere is heavier, almost certainly a result of the concentration
of more and larger yolk granules in the vegetal hemisphere. Therefore, if the
egg rotates to a new orientation with the yolk down and the animal hemisphere
up, it must be the case that at this point the yolk and other egg contents are
not free to be redistributed within the egg but are secured in place. The animal–
vegetal axis now marks the future anteroposterior axis of the embryo.

A number of important events occur at fertilization, although they seem to happen quite differently in different groups of amphibians. In Anura there is entry by a single sperm and this sets in train a new organization in the egg and the establishment of a new axis of symmetry—the future dorsoventral axis. Recent studies (e.g., Ubbels, Hara, Koster and Kirschner, 1983) show that at fertilization, in connection with an "activation wave" starting from the sperm entry point, cortical granules are extruded from the egg, the pigment cap contracts, the vitelline membrane separates, and the egg rotates. The pigment cap then relaxes again. Then a postfertilization wave starts from the same point, coinciding with reorganization of the cytoplasm under the influence of the sperm centriole and a newly formed spermaster. A central region of yolk-free cytoplasm forms and then moves to the future dorsal side of the cortex as the "dorsal cytoplasm." In anurans this dorsal region is defined opposite the sperm entry point and is marked superficially by the formation of the "gray crescent" in *Rana*. In *Xenopus*, pigmentation of the egg is less pronounced and no "gray" crescentic mark is formed, although the special properties of this region are nonetheless manifest.

The gray crescent or its equivalent is important because it represents the first step in the process of continued asymmetrization of the embryo, marking the future dorsal side of the embryo. But, as is well known, it also marks the site of the dorsal lip of the blastopore, the site of gastrulation. Throughout the blastula stage, presence of this special region is essential to the flow of developmental changes. Therefore, its formation is one of the first major reorganizations within the egg, a specialization that will form a determining feature for the whole of early pattern formation. The gray crescent region is formed as a result of relative movement between the cortical and deep regions of the egg (Vincent, Oster, and Gerhart, 1986). The cortex moves about 30 degrees in one direction toward the side of the egg at which the sperm entered, pulling the deeper part of the egg contents up on the gray crescent side. Where a gray-colored patch appears, it seems to be due to a resultant thinning of pigmented zone at this equatorial region (Frank et al., 1983; Elinson, 1980; Gerhart, Black, and Scharf, 1983). This reorganization is complete before the first mitotic division occurs.

From this point onward, all cell behavior in the future dorsal region is subtly or overtly different from the rest of the embryo. The dorsal region acts as a major determinant, and yet it appears to be based on an extremely simple, even crude, reorganization within the egg, based on new juxtapositions of regions of the cortical and deeper regions (Gerhart, Black, and Scharf, 1983). Relative movements within the egg seem to involve the cytoskeleton. They can be blocked by UV radiation, by cold and pressure, and by chemical agents that depolymerize microtubules. When this happens, the dorsal axis fails to materialize and gastrulation occurs around the whole equator of the egg (Malacinski et al., 1980), forming an essentially "ventral embryo."

If the egg is physically prevented from rotating within the vitelline membrane within a critical time before normal gray crescent formation, the egg contents will reorganize under gravity alone and a gray crescent region will form on the uppermost side of the egg, even if that is not the point opposite the sperm entry point. Thus the sperm entry point is important in triggering events normally, but is not itself determinative of gray crescent formation (Kirschner et al., 1981; Gerhart et al., Gerhart, Black, and Scharf, 1983). This has also been shown in experiments in which the egg is artificially caused to reorganize through being held in a rotated position or through mild centrifugation (Pasteels, 1948; Gerhart et al., 1981; Black and Gerhart, 1985). Ubbels et al. (1983) show that in such cases, although the egg contents reorganize and a vitelline wall forms in the correct relationship to the future dorsal region, the clear zone of dorsal cytoplasm and the pronuclei do not move, presumably being held in place by the cytoskeleton.

Development can also proceed without fertilization, if the eggs are merely pricked (Ubbels, 1977), and in eggs that are cultured in vinblastine, which prevents microtubule assembly. In these cases there is no postfertilization wave and no sperm centriole. The gray crescent forms later than usual and not always opposite the point of pricking. Eggs that have been blocked by UV radiation or other treatments can be "rescued" by means of artificial rotation (Scharf and Gerhart, 1980; Gerhart, Black, and Scharf, 1983). Further, normal fertilized eggs can be reorganized by centrifugation to form a second gray crescent on the opposite side, resulting in a double-headed embryo (Black and Gerhart, 1985; cf. Pasteels, 1948).

It seems that the organization of the egg involves strict distribution and localization of a number of factors (and not just the yolk) that are normally arranged in an array connected with the spatial organization of the cytoskeleton. Neff et al. (1984) have proposed a "density compartment model" based on major and minor compartments within the egg that help control cytoplasmic rearrangements (see also Neff, Malacinski, and Chung, 1985, for a discussion of models of how the reorganization is driven). In the period immediately after fertilization, when the definitive cytoskeleton is still being assembled, there is a phase during which the organization of the egg can be continued under the guiding influence of the distribution of the yolk and other cytoplasmic factors by gravity. Gradually, however, the cytoskeleton takes over a fully controlling role, demonstrated by the fact that recovery from rotation and other experimental insults is not possible after the cytoskeleton is more complete (marked roughly by the time of gray crescent formation). The organization of the fertilized egg is therefore a dynamic process in which cytoplasmic and cytoskeletal factors are major components but not in a strictly deterministic sense. In particular, the dorsoventral organization of the egg results from interactions among these components in a way that can be shown by experiment to be quite fluid. This is therefore a demonstration of the fact that the normal development of an embryo, where the various regions appear to show fixed fates, is not always a reliable guide to the potentials inherent in the causal mechanisms underlying development. Finally, notice should be made of the interesting observation that gray crescent formation can be induced precociously in nonactivated *Ambystoma* oocytes by use of inhibitors of protein synthesis (Gautier and Beetschen, 1985), but not in enucleated eggs. Enucleated eggs can, however, be rescued by injection of nucleoplasm from a normal oocyte, which seems to contain a protein synthesis inhibitor.

CLEAVAGE AND THE BLASTULA

The planes of the first cleavages are extremely important in amphibian development. The first cleavage plane seems to be defined, in anurans at least, by the sperm entry point. The first cleavage separates the gray crescent region exactly in two and produces two equal blastomeres with a complement of animal gray crescent, and vegetal material. As we have already noted, the first two blastomeres can be separated and will grow up to normal individuals (see Chapter 4; and Cooke and Weber, 1985).

Within the developing blastula there are at first three main regions: the animal cap and the vegetal region, with the gray crescent region occupying a position at the equatorial boundary between the two, on the dorsal side. Within the blastula there then occurs an inductive process by which the marginal part of the animal cap, under the inductive influence of the vegetal region, forms an intermediate zone that will give rise to the mesoderm of the future embryo (Dale, Smith, and Slack, 1985). If the vegetal region is damaged by UV radiation, this induction is prevented. Thus formation of the future mesoderm region seems to occur through what Nakamura (1978) calls "regulation of a gradient": a strong vegetal–animal gradient arising from and controlled by a zone in the vegetal region that has become established at or before fertilization (see also Ubbels et al., 1983). At this point, the blastula has been organized into three presumptive zones: ectoderm (animal region), endoderm (vegetal region), and an equatorial "marginal zone" of mesoderm. The old gray crescent is in the dorsal marginal zone, and it is at this point that the dorsal lip of the blastopore (the primary organizer) will soon emerge.

Experiments by Nakamura and Matsuzawa (1967) and Nakamura and Tasakai (1970; see review in Nakamura and Toivonen, 1978) have demonstrated that by about the 64-cell stage, cultured explants from the animal cap will produce ciliated epidermis; the marginal zone will produce notochord, muscle, and neuroepithelial cells; and explants from the vegetal region will produce erythrocytes and epidermis. This indicates that these regions are in a general sense already partially committed as well as specified, and this commitment must mark the activation of the zygotic genome.

The regulative powers of the blastula are extremely interesting. We have already noted the importance of the gray crescent of the egg and its relationship to the future dorsal lip of the blastopore. Curtis (1960) showed that if the gray crescent region of the cortex of the fertilized egg were excised, the embryo would not develop. However, in the 8-cell blastula, if the same region is excised, the embryo can reproduce all the missing structures and gastrulation proceeds normally (Curtis, 1962). That the excised region is indeed that region essential for future gastrulation seemed to be shown by parallel experiments in which an explant from that region is implanted to the opposite side of an otherwise normal fertilized egg, in which case a double embryo is formed because of the formation of a second dorsal blastopore lip. Finally, a similar attempt to produce a double embryo by implantation of a second dorsal marginal zone in an 8-cell blastula failed to produce a normal embryo (so that the potential effects of the second dorsal marginal zone were regulated away by the embryo). These experiments have been the subject of much criticism (Kirschner and Gerhart, 1981; Malacinski, Chung, and Ashima, 1980). Some of the results on the egg stage appear to be due to the effects of rotation of the egg in the procedures. Nakamura repeated the 8-cell experiment at the 32-cell stage with the same result. Neiuwkoop (1973) experimented with combinations of different regions of the blastula and reached similar conclusions. If isolated portions from the animal and vegetal regions were combined, the embryo could regulate formation of a marginal zone and gastrulation could proceed. Artificial embryos consisting of vegetal or animal material alone could not develop. Thus regulation only occurs as a result of functions depending on at least a gradient and probably upon some more complex basis of interaction across the embryo.

Similar results are shown in the experiments (Chapter 4) in which partial embryos are cultured, having been divided at the four-cell or eight-cell stage. Normal development only occurs when the right proportions of animal and vegetal material are present (Spemann, 1938; Nakamura, 1978; Kageura and Yamana, 1983).

By the late-blastula stage a quite detailed fate map of the embryo can be drawn and, as we have just seen, this specification is accompanied by a degree of commitment. However, commitment of most tissues does not become complete until gastrulation.

GASTRULATION

The mechanical aspects of gastrulation are familiar from any textbook of embryology. What may not be so clear is that gastrulation does not always occur in the wonderfully simple textbook fashion (which is based on urodele development). In many anurans the mesoderm delaminates at the blastula stage.

Mechanical reorganization of the whole embryo at gastrulation depends on subtle differences in cell properties of different regions of the blastula and on the response of cells to extracellular matrices laid down in the embryo (Figure 4). The presumptive tissue regions that can be traced through classical fate map markings, using vital stains or other agencies (Cleine and Slack, 1985), can also be characterized at the electron microscope level in terms of differences in cell shape and behavior. Gastrulation has been an obvious subject for computer modeling experiments in which relatively few parameters of cell shape and growth are involved (e.g., Odell et al., 1981).

It is starting to become obvious that a very important part of the process of gastrulation is a process of response of cells not only to each other but also to a strictly defined macromolecular extracellular "skeleton." A body of results already exists with respect to the cell surface glycoprotein fibronectin (Thiery et al., 1982; Doubard and Thiery, 1982; Boucaut and Darribiere, 1983; Nakatsuji, Smolira, and Wylie, 1985). Boucaut et al. (1984) show that prevention of fibronectin synthesis in the late-blastula stage, prevents gastrulation, but, interestingly, neurulation is not disturbed. Nakatsuji and Johnson (1983; cf.

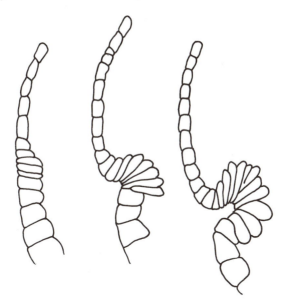

Figure 4 Change in cell shape at the dorsal lip of the blastope in amphibian gastrulation. After Holtfreter and Keller.

Boucaut and Darribiere, 1983) show that during gastrulation itself the cells of the presumptive mesoderm migrate along the inner side of the ectoderm following a pattern of fibronectin fibrils. These fibrils are absent from the endoderm and mesoderm cells.

The essence of gastrulation is primary induction by the "organizer," the dorsal lip of the blastopore, as the classic experiments of Spemann and Mangold demonstrated some 60 years ago. But despite the obvious importance of primary induction, both in terms of its morphogenetic functions and as a supreme example of a set of inductive interactions within an embryo, there remain almost as many questions unsolved today as there were originally. Specifically, it is not clear how the organizer is itself organized and how the effects of the primary organizer are caused—the actual mechanism of induction. What is clear about primary induction is that as cell groups pass over the dorsal lip of the blastopore they become more fully committed. The effect of primary induction is to add a major layer of irreversible change to the process of activation of the zygotic genome. For example, the experiments of King and Briggs (1956) on nuclear transplantation (see page 34) showed that while nuclei from the blastula were capable of supporting normal development in an enucleated egg, nuclei from cells that had undergone gastrulation became largely incapable of thus functioning.

Parallel experiments on the culturing of tissues to test for their ability to autodifferentiate show that, after having passed over the dorsal lip of the blastopore, most of the presumptive tissue regions have become quite fully committed, as far as the definition of class properties. There is also a most inter-

esting set of experiments in which the organizer itself is tested for time changes
in its properties. In this it is found that the dorsal lip from an early stage of
gastrulation, when cultured with presumptive neurectoderm from the late-blastula stage, will form structures characteristic of the forebrain, midstage dorsal
lip will form midbrain structures, and late-stage dorsal lip will only form hindbrain structures—corresponding exactly with the sequence in which the prechordal plate and chordamesoderm are put into place underneath the neurectoderm and induce the formation of the brain in an anteroposterior sequence.

As a result of these and similar experiments, models have gradually evolved
that explain the regionalization of the embryo according to multiple gradient
systems. Toivonen (1978) reviewed research on the two-gradient model of gastrulation. There is postulated "the existence of two active principles which both
form gradients on the dorsal side of the developing embryos: the neuralising
(N) principle, which is strongest in the dorsal midline and forms a decreasing
gradient laterally and also somewhat caudally, and the mesodermalising (M)
principle which is strongest in [the] caudal midline and decreases cranially and
laterally." Slack (1983) defined a similar model for the blastula with gradients
in the animal–vegetal and ventrodorsal axes. He assigns codings 1–4 along
each axis and this delimits the blastula into a series of zones with different
scores. The ability of an artificially manipulated embryo to regulate the formation of new gradients when regions of the embryo are excised will depend
on the scorings of those parts that are lost or remain (Figure 5). He notes that
in this scheme, the region with the highest scores in both axes is the region of
the future organizer.

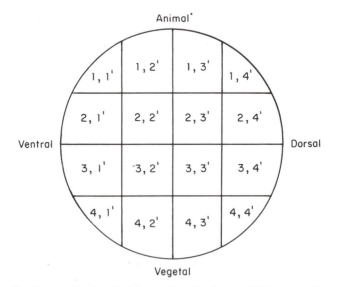

Figure 5   Slack's model of a double gradient in the amphibian egg. Each region is
coded for its position on two gradients. The highest score is in the region that marks
the future dorsal lip of the blastopore. Redrawn from Slack (1983).

GASTRULATION TO NEURULATION

Gastrulation forms an embryo that is completely enclosed in presumptive ectoderm, with the neurectoderm placed down the dorsal midline. The archenteric space is surrounded by chordamesoderm dorsally and endoderm ventrally. There then follows a cascade of inductive interactions. The chordamesoderm induces the overlying neurectoderm to form the neural tube. This occurs as a classic case of folding of an epithelium (Schoenwolf and Franks, 1984; Tuckett and Morriss-Kay, 1985) apparently driven by changes in cell shape and preferential orientation ofthe axes of cell division. A. G. Jacobson et al. (1986) have recently modeled these processes in terms of the cortical tractor model of Oster et al. (1983). The neural tube in turn induces formation of the notochord from the central stripe of chordamesoderm, while the more lateral regions form the paraxial and lateral plate mesoderm. Meanwhile, the endoderm grows around to enclose the archenteric cavity, leaving the mesoderm in the body cavity. As the neural tube forms, the neural crest is induced (see next chapter). Thus in a very short time, all the major features of vertebrate organization have been established in the gastrula, and the stage is set for the morphogenetic phases of late pattern formation to begin.

The work briefly summarized above suggests that there is a complex relationship between determinism and regulation in the early development of amphibians that involves a remarkable "redundancy." The results of the rotation experiments on the fertilized egg, first conducted by Pasteels and more recently brilliantly extended, show that formation of the gray crescent region is caused in normal eggs by cortical changes activated by the sperm centriole. However, these can be overridden in abnormally rotated eggs by the action of gravity alone. Comparison of changes in egg organization under normal and experimental conditions shows that a major factor in the formation of the dorsoventral axis of the egg is movement of the yolk and other cytoplasmic "compartments" and formation of the "vitelline wall." Once these changes have occurred, and the dorsoventral axis is fixed, the egg is strictly organized in such a way that the first cleavages will proceed in the correct relationship to the dorsoventral and animal–vegetal axes of the egg. This organization of the egg may or may not be the result of regulation of gradients within the egg, but the result seems to be that the gray crescent region comes to contain causal cytoplasmic determinants (which may possibly be transplantable from the eight-cell stage back to the egg).

Once cleavage has proceeded to the 8-cell stage, the specific role of the gray crescent seems to have changed again. Explanation of this region will not produce a double embryo. Loss of this region can be regulated for. A gray crescent region can be recreated in embryos formed from artificial combinations of animal and vegetal pole cells alone. Gimlich and Gerhart (1984) showed that 64-cell embryos that had been irradiated with UV before fertilization, and thus were prevented from dorsalizing (lacked a gray crescent), could be rescued by implantation of one to three vegetal blastomeres. Formation of the organizer, although it is topographically in the same place, is therefore potentially independent of the causal mechanisms involved in the oocytic and 1-cell stage factors that produce the gray crescent itself. There is a new set of causal mechanisms, again centered upon the dorsal side of the dorsoventral axis, which control formation of the organizer, and their nature is currently unknown. The processes of gastrulation produce the first serious irreversible commitments in many cell types. These cells thus acquire new properties that quickly become the basis of inductive signaling (often mutually interactive), which causes a cascade of new levels of commitment and new interactions).

Early development in amphibians is therefore both regulative and determinative, and there is something of a cyclic alternation between the two. An important aspect of this cyclic scheme of causal mechanisms is that at each stage where new controls are created there is possibility for change in both the controls and the patterns they cause (see, for example, artificial creation of macrocephaly in frogs by inhibition of cellular rearrangements in gastrulation by Kao and Elinson, 1985). At the same time, the fact that, at least in the case of formation of the dorsoventral axis of the egg, there is redundancy in the causal mechanism, provides an opportunity for these processes to be buffered against such changes. Because control is not achieved via the operation of strictly determined causal agencies, the sequence of controlling mechanisms provides a flexibility that in the short run benefits the species by providing buffering against genetic or environmentally induced mistakes and in the long run provides the potential for developmental plasticity on which evolutionary change depends.

6

Later Pattern Formation: Morphogenesis

The early phases of pattern formation, as described in the two previous chapters, set the stage for all that follows—for the whole grand sweep of morphogenesis by which the phenotype is created. Right from the oocyte stage there is an ordering of the embryo, both in the sense of spatial patterning and in the sense of setting in place the components of developmental processes. As pattern formation continues, particularly in "regulative" embryos, the ordering of the embryo becomes more and more specific. A point is reached that is quite impossible to define but nonetheless real, when the embryo is set up in such a way that all the components are in place. At least, everything has been specified. Such an embryo, for example the amphibian neurula, may look very little like the final phenotype, but from this point onward morphogenesis represents a working out of potentials that have already been established. All the basic morphogenetic information is in place.

Morphogenesis is both simple and complex. It is simple because relatively few processes will be involved. It is complex because of the diversity of cell and tissue types that is involved, because of the subtlety of control of differentiation and even size and shape in organogenesis, and because of the complexity of both embryonic and subsequent adult function. Finally, it is complex because of the interactivity and "wholeness" of the developing embryo as well as the multiplicity of the parts making up that whole. It is with the simplicity of morphogenetic processes, the relatively small number of cell and tissue-level processes involved (processes that are common to all morphogenetic systems), that we will be concerned first. In order to understand both the basic rules of development and the way in which they relate to mechanisms for the introduction of evolutionarily significant phenotypic variation (Chapter 2), we must understand first these common processes of morphogenesis, and the way in which they operate under the rules and mechanisms of pattern control discussed in the previous chapters.

Morphogenesis involves a relatively small number of phenomena characteristic of all cells and tissues, and their relation to features of the extracellular environment. It is not necessary to rehearse here all the basic features, but it is necessary briefly to note some of these processes before proceeding to some examples of developmental/evolutionary problems.

1. *Cellular Phenomena*. The first stages of pattern formation place embryonic cells in a particular position within the embryo. From this point they will divide and produce daughter cells whose various histories and fates will constitute morphogenesis. As Wessells (1982) has emphasized, a major feature of the constituent cells of multicellular organisms is that they are only capable of undertaking these various histories while in contact with a solid substrate— either contact with other cells or with some cellular product such as an extracellular matrix (ECM) and mediated by the fascinating class of molecules known cell adhesion molecules (CAM).

Probably most cells reach their final appointed positions within the embryos largely by passive means (passive, that is, as far as the individual cell is concerned), by the growth of a mass of mesenchyme, or by the growth and folding of an epithelium. There are, however, many cases where cells migrate as groups or individuals, often traversing relatively large distances to reach their final destination, where they settle and differentiate. Classic among these migrating cells are the neural crest and placodal cells of vertebrates. Study of such cells is not only of vital importance to the understanding of processes of embryonic development, but also to the unraveling of those processes by which normal cells may produce cancerous lineages that also migrate and invade other tissues.

The migration of cells involves both the motility of the cells themselves and the nature of the extracellular environment through which they move. The extracellular spaces in the embryos are usually rich in macromolecules specifically produced by the surrounding cells to create a particular extracellular environment. The pathways along which cells migrate are especially rich in collagens, proteoglycans such as fibronectin, and hyaluronic acid (see extensive reviews in Hay, 1983; Kemp and Hinchliffe, 1984; Trelstadt, 1984). Mention has already been made in the previous chapter of the complex role of fibronectin in the amphibian gastrula. Embryonic cells migrate through their own movements, but it is also worth noting that to a certain extent the relative positions of cells may change not just by their own migration, but by the mass movement of the environment past them. In either case, a relative interaction of cell and environment is required.

One of the most interesting and well-studied examples of cell migration in vertebrates is that of the neural crest (see below). Crest cells migrate in pathways that are strictly defined both by the gross architecture of the extracellular spaces and by the ECMs contained within them (and therefore by the tissues that lay down the ECMs). Neural crest particularly follows pathways rich in fibronectin produced by the basement membranes of epithelia. The direction of cell migration is controlled by these tracts and pathways, by a general high- to low-density trend, and by intercell contact (Goodday and Thorogood, 1985; Erickson, 1985). There is also evidence to suggest that the neural crest cells take up precursors of macromolecules from the pathway; this raises an interesting question for the cells that follow behind and therefore for the control over overall cell migration (Brauer et al., 1983). There are many unexplained features of these cell migrations, not the least being that serum-coated latex microspheres will apparently migrate by themselves down the pathways, unattached to cells (Bronner-Fraser, 1982, 1986). Fibronectin-coated spheres fail to migrate. Injected latex microspheres thus offer a way of visualizing migratory pathways and the extent of ECMs (Meier and Drake, 1984).

Obviously an equally important factor in morphogenesis will be the division of embryonic cells to produce lineages. The rate of cell division will control the relative size and in part also the shape of the cell mass so produced, mesenchyme or epithelium. As we will note in Chapter 7, cell density and cell number are important elements in the operation of pattern control mechanisms in development. The orientation of each dividing cell will play an important role in controlling the shape of the resulting population. The most obvious case of this is in spiral cleavage, but it is also crucial in epithelial morphogenesis, such as neurulation. The intrinsic and extrinsic control of rates of cell division must be highly significant in terms of production of new phenotypic variation.

From a surprisingly early stage in development, cell death in specific regions of the embryo provides an important mechanism for shaping and forming organs. Cell death is an important feature in such disparate systems as the patterning of the nervous system of insects (Loer, Steeves, and Goodman, 1983) and the shaping of the digits in the tetrapod limb (e.g., Fallon and Cameron, 1977; Hinchliffe and Thorogood, 1974). Once again, the question of intrinsic versus extrinsic controls arises.

While it is possible to catalogue phenomena in the biology of single embryonic cells that contribute to the processes of morphogenesis, the fact remains that even migrating cells such as the neural crest cells of vertebrates do not really act alone. The contact dependence of cells on an ECM provides a strong measure of organismal integration. The contact-dependent switching on and off of cell motility, and contact guidance, provide even stronger integration. And perhaps nowhere is such cell-to-cell contact more important than in mechanisms of induction.

2. *Tissue-Level Phenomena.* As already noted, there are two basic patterns of tissues that are important in morphogenesis: epithelia and mesenchyme. A epithelium is always a single layer of cells, each cuboid in form with strong cell-to-cell contact. The crucial feature of all epithelia is the presence of an extracellular basement membrane. This basement membrane both integrates the functions of the epithelial cells and provides a major substrate for other cells in the embryo.

Because of their discrete spatial ordering, epithelia have certain unique properties. As noted in Chapter 2, the behavior of an epithelium is something like the behavior of an elastic sheet. It can be folded and distorted in various ways, but the basic sheetlike properties impose important constraints on the sorts of structural configurations that can be produced by epithelia. These simple properties also make epithelia readily accessible to theoretical modeling (Odell et al., 1981; Oster et al., 1983).

One thing that epithelia cannot do very well is produce a solid mass of cells, such as the blastema of a limb element. Epithelia can readily delaminate cells from their surface, although in order for this to happen the basement membrane has to be disrupted. When an embryo "wants to make" a solid structure, most often a mesenchyme of cells is the first agency. The other thing that epithelial cells usually cannot do is migrate. Migratory cells in embryos are almost invariably mesenchymal.

Mesenchymes differ from epithelia in several significant ways, always being multidimensional rather than one cell thick, and lacking a basement membrane. Whereas in vertebrates, ectoderm and endoderm are always epithelial, the mesoderm is the major source of mesenchyme. However, the neural crest cells of vertebrates, so often mentioned here, are the major exception to this rule— forming an "ectomesenchyme" of migratory cells. (Incidentally, the discovery of the ectomesenchyme of vertebrates by Platt, 1893, and others, was the major crack in the dominant germ-layer theory of nineteenth-century embryology according to which all Metazoa were created from three embryonic cell layers: ectoderm, mesoderm, and endoderm.)

The clumps of mesenchyme cells that form organ rudiments may arise through migration to a particular site and/or by condensation from within an homogeneous cell mass in the relevant position. Blastemata may be shaped through further cell divisions, through recruitment of cells from neighboring mesenchyme or migrating cells, through change of cell shape, through effects caused by ECM, and through cell death either within the blastema or in the surrounding tissues (see below). The ways in which such processes are controlled are still very little understood. It is likely, however, as already mentioned, that ECMs are important, especially in the formation of initial rudiments and in controlling cell migration and cell-to-cell interactions (Oster, Murray, and Harris, 1983).

Mesenchymal cell masses can readily form epithelia by daughter cells that form in plates and develop strong cell contacts. Otherwise, mesenchymal cell masses tend to consist of rather more separated cells. The ECM may be important in keeping the mesenchymal mass together, just as it is important in guiding the migration of ectomesenchymal cells and their later clumping to form tissue blastema (see below).

3. *Epitheliomesenchymal Interactions.* In vertebrates, at least, and perhaps to a varying extent in invertebrate animals, structures of mesenchymal origin only form as a result of highly specific interactive processes involving mesenchyme and an epithelium (review: Sawyer and Fallon, 1983). An obvious example is the formation of teeth. Very roughly, mesenchyme (ectomesenchyme, in fact) in the jaws forms a dental lamina in which groups of dental papillae assemble. The condensations probably occur as a result of an inductive interaction between the ectomesenchyme and jaw mesoderm. Then the papilla induces the formation of a specialized dental epithelium in the overlying ectodermal epithelium. This becomes folded over the tooth papilla so that it is actually double, with an inner and outer portion. The inner dental epithelium then reinduces the tooth papillae to start to differentiate. The ectomesenchyme forms odontoblasts, which start to secrete the protein matrix of the dentine. The odontoblasts in turn induce the inner dental epithelium to form enamel

over the cap of the tooth. The shape of the tooth is determined by the mesenchymal component, and the causal factors probably include mechanical pressures developed inside the papillae as well as the complex folding of the inner dental epithelium.

By now the experiments of Kollar and Fisher (1980) are famous, though not totally confirmed. They combined chick oral epithelium and mouse mesenchyme in vitro and produced molariform teeth. Thus, although birds have been unable to develop teeth for some 50 million years, their ectoderm has retained its inductive capacity vis-à-vis dental mesenchyme. Similar results have been obtained with crocodile–mouse hybrids, and in fact many other kinds of chimeras involving epithelia and mesenchyme have been produced in vitro, including many on hair and gland morphogenesis (review in Slavkin et al., 1984). All show that in such combinations, the essential characteristics of the resultant organ follow the mesenchymal component rather than the epithelial portion.

The properties of epithelia and mesenchyme, and particularly their interactions, offer some obvious vehicles for the generation of evolutionarily interesting variations. The simplest would be the failure of an organ rudiment to form because of a shift in timing causing the required epitheliomesenchymal interaction to fail. Perhaps this was involved in the loss of teeth in birds. Relatively simple changes in external conditions surrounding mesenchyme migration or epithelial folding will have a major effect on morphogenesis, as will changes in the rates of cell division within, or recruitment to, mesenchymal blastemata. Oster and Alberch (1982) attempted, for example, to show that a whole range of integumentary organs in vertebrates—hair, teeth, scales, glands— can be related as modulations of a single scheme of epitheliomesenchymal interaction involving the dermis.

THE TETRAPOD LIMB

The development of the tetrapod limb provides an excellent first example by means of which to examine the processes and patterns of morphogenesis, in this case in a single, albeit complex organ. Limb development has been studied extremely thoroughly both in amphibians and in chicks. (By contrast, the development of the paired fins of fishes has been greatly neglected.) There is no purpose in repeating here all the details of descriptive and experimental embryology of the limb that are available in the textbooks. The reader is referred particularly to the elegant summary by Hinchliffe and Johnson (1980) for a review of the subject. Instead we will concentrate on certain features of pattern formation in the limb that serve to highlight the question of the origin of evolutionary novelties and its relationship to the mechanisms of limb development.

In all vertebrates, the limb is organized first as a simple limb field in the flank. An important feature of this is the formation of a "Wolffian ridge" of epidermis along either flank. The ectodermal covering of the central part of the ridge becomes developed as the apical ectodermal ridge (AER), which has a major role in pattern formation in the limb. Mesenchyme accumulates within the fold, setting in place the components for epitheliomesenchymal interactions, and the basic limb bud is now in place. Very quickly dorsoventral and anteroposterior asymmetries start to appear. The limb bud changes shape to form a discrete projection from the body, the proximal part becoming slightly pinched off, and a distal footplate broadening out where the digits will form.

The origin of the limb bud mesenchyme is complex. The limb bud always includes mesoderm of somitic origin, from both the somites and the lateral plate, but no ectomesenchyme. In chicks, as shown by the experiments of Christ, Jacob, and Jacob (1977), the somites contribute the presumptive myoblasts of the muscles. Presumptive chondroblasts and fibroblasts derive from the somatopleural mesoderm. The innervation of the limb bud comes relatively late in development, axons invading from the relevant segmental position of the dorsal nerve cord. The limb buds, pectoral and pelvic, therefore bear a relationship to the body segments. But interestingly, the number of segments involved and which segments, may vary among different species, which is evidence of a general phenomenon that is termed developmental plasticity (where plasticity is perhaps another word for uncertainty).

The results of much experimental work has produced a general hypothesis: the Saunders–Zwillig hypothesis of limb morphogenesis. According to this hypothesis, the AER is maintained by a factor from the mesoderm (the apical ectodermal maintenance factor, or AEMF) which forms a proximodistal gradient in the limb bud. The general characteristics of the limb are controlled by the mesenchyme itself, although the ectoderm, for a short period of time, controls the dorsoventral polarity (Patou, 1977). Very early in development, a second focus of pattern formation develops. This is the zone of polarizing activity (ZPA), which controls anteroposterior polarity and differentiation in the limb bud. Wolpert and co-workers have developed a "progress zone" model to account for the formation of the limb in a proximodistal sequence. In this model, as mesenchyme is added to the limb bud distally and leaves the influence of the AER, it becomes committed. The experimental basis of this hypothesis is, again, summarized in Hinchliffe and Johnson (1980).

As development proceeds, the various muscle and skeletal blastemata differentiate out of the mesodermal core of the limb bud. It is worth emphasizing that studies of the development of both the chick and amphibian limbs (Hinchliffe and Johnson, 1980; Hinchliffe and Griffiths, 1983) show no sign of any recapitulation in the pattern and sequence of development of the skeleton. In general the blastemata are laid down in just that pattern that is characteristic of the adult limb. There is, however, a significant exception in chick in that a precartilaginous ulnare is formed in the wing and then does not develop further.

Comparative morphological evidence had demonstrated since before the time of Darwin that all tetrapod limbs are constructed according to a single homologous pattern, and there is general agreement that this limb type arose only once in evolution. This means that all the different limb phenotypes that have been expressed during tetrapod diversification must represent modulations of a single developmental theme of pattern formation. However, there has been some debate concerning the details of the basic phenotypic pattern itself and much speculation concerning the underlying morphogenetic processes. All attempts to derive a generating mechanism for limb patterns have had to start with the common pattern drawn from morphology (thus showing the value of the comparative approach). But it is only very recently that a really persuasive attempt to describe that pattern has been achieved. Most authors have taken the basic pattern of the limb to be a symmetrically branching system: the single proximal element (humerus or femur) forms an axis that branches, giving two elements (radius and ulna or tibia and fibula, respectively). Then, each of these more distal elements supports a branching series. Shubin and Alberch (1986) show that instead the comparative data really support an asymmetric scheme in which there is a pair of branches stemming from the single proximal element (humerus or femur), but only the postaxial stem branches (Figure 6).

Various theoretical schemes, some modeled by computer, have been drawn up with the aim of demonstrating how variant limb phenotypes could be generated by varying relatively simple parameters. One of the most interesting of these models, even though it is wrong in details, is that of Goodwin and Trainor (1983). Their model is one in which the positions at which mesenchymal blastemata of the skeletal elements form nodal points in a field controlled by a set of gradients (purely hypothetical in nature). One obvious drawback to their model is that it ignores the possibility of local interactive effects in the limb bud and phenomena such as division or fusion of blastemata. One of the great advantages of this model is that it concentrates our attention on the fact that the blastemata are the products of the pattern-generating mechanism, not the prime components of it. The pattern of blastemata essentially falls out according to whole-limb properties created by a field of gradients. Only the pattern-generating mechanism itself is real and passed on from generation, not the blastemata. While this might seem a trivial semantic point, if this view is correct, it poses serious problems for homology of elements. We have little hesitation in calling the stylopodium or zeugopodium of all tetrapods limbs homologous. But it may be impossible to say that a given carpal or tarsal element of limb A is homologous with an element limb B. In a three-digit limb it may not really be the case that specific digits are missing: rather, the three-digit condition may be homologous with the five-digit condition in another taxon, even in evolutionary transitions where we can see transitional phases from taxon to taxon. Each taxon only shows one condition at a time, after all, and all share only the common field phenomena. This is a viewpoint that needs much careful thought (see, for example, Roth, 1987).

Shubin and Albrech's model (1986) attempts to include both "global and local interaction of morphogenesis" in producing three basic phenomena of cartilage rudiment formation: de novo appearance, branching, and segmentation of prechondrogenic blastemata. The model is linked to morphogenetic models of mesenchymal blastemata by Oster, Murray, and Maini (1985), relating mesenchymal cell behavior to properties of the ECM and the empirical observations

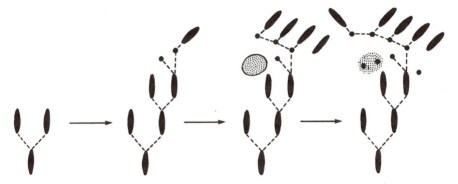

Figure 6 Shubin and Alberch's model of the sequence and pattern of blastemata formation in the forelimb of a tetrapod—the turtle *Chelydra*. Redrawn from Shubin and Alberch (1986).

of chondroblast behavior described by, for example, Archer, Hornbrach, and Wolpert (1983), Archer, Rooney, and Cottrill (1985), and Zanetti and Solursh (1986).

We are still quite a long way from having enough information about morphogenetic processes in the tetrapod limb bud to set out a complete account of the mechanisms controlling the emergence of the phenotype. Thus we cannot, except in the most general terms, discuss how new phenotypes could be created. We know that the limb bud contains mesenchyme from two different sources. These mesenchyme cells form a superficially homogeneous mass in the limb bud, but it is clear that the presumptive fates of the cell types affect their differential behavior right from the start. Thus mesenchyme is not totally naive when first incorporated into the limb bud. However, the AER has a major controlling and regulative role in addition, strongly mediated through ECMs. Experimental work in vitro shows that the behavior of limb bud mesenchyme depends in part on its intrinsic patterning and in part on external factors. Archer et al. (1985) show, for example, that cell density plays a major role in determining whether mesenchyme would form cartilage sheets or nodules. Solursh (1984a, 1984b) has demonstrated the role of the ectoderm in the patterning of chondrogenesis.

Prechondrogenic and premuscular mesenchyme in the limb bud must react separately but at the same time to influences operating within the limb bud. One of the first causal factors is the establishment of ECMs rich in hyaluronate by the limb bud epithelium. The ECM must be the major site of early expression of pattern information created by the AER and ZPA. The next stage is that, under the control of the ectoderm and mediated through these ECMs, patterns of mesenchyme condensation appear and in these prechondrogenic condensations there is secretion of high concentrations of glycosaminoglycans (see, for example, Kosher et al., 1986).

The full role of the ectoderm at various stages of limb development is still unclear. For example, Martin and Lewis (1986) found normal development of chick limb is the absence of dorsal ectoderm removed at stage 19. They conclude that the major patterning role must reside in the AER. However, Hurle and Ganon (1986) found that if dorsal ectoderm were surgically removed from the chick limb at stage 23, interdigital mesenchyme would become chondrogenic.

The behavior of cells in the preskeletal condensations seems to be a function in part of differential cell density. Those chondroblasts in the center where cell density is highest are rounded in shape and will become arranged in columns. Those on the periphery are more flattened, and they become arranged around what will be the outer surface of the blastema; they will form the perichondrium. Oster et al. (1985) have proposed a model in which the mesenchyme condensations arise through a sort of feedback between the mesenchyme and the ECM. The chondroblasts in the condensations produce hyalouronidase, which degrades hyaluronate. This causes "deswelling" of the hyaluronate, which is a hydrophilic polymer. This brings the cell closer together, allowing the formation of intercellular junctions and intercellular cell tractions that further bring the cells together.

The formation of a rudimentary perichondrium in the form of flattened chondroblasts around the rudiment physically constrains its future growth. Archer et al. (1983; 1985) show that the blastema remains open-ended at the distal end, and here growth can occur through recruitment of new mesenchyme cells. Shubin and Alberch argue that this recruitment can then lead to segmentation of branching of the blastema. As the chondroblasts mature into chondrocytes and the cartilaginous element starts to acquire its distinctive shape, the chondrocytes begin a new round of matrix secretions—both of the cartilage matrix itself and of the ECMs around the condensation (see Thorogood, 1983; Shinomura et al., 1984).

Morphogenesis of muscle blastemata is more complex, if only because the fibroblasts forming the connective tissue and the myoblasts are of different origin. But the general principles are the same. Chevalier (1978) and Jacob and Christ (1980) show that splitting of the premuscular mesenchyme is largely controlled by the connective tissue.

The role of the vascular system in the patterning of blastemata is little understood, although it seems likely that if there is a vascular trace between two blastemata the possibility of their subsequent fusion is diminished. The vascular system in the limb bud starts out as a relatively diffuse capillary network and during morphogenesis the definitive circulation is produced, presumably interactively with the morphogenesis of the rest of the limb. Innervation is the last part of the process and seems to have no direct role in shaping the blastemata. The innervation of the limb comes from invading axons from the trunk, the segmentally arranged nerves matching up with mesodermal blastemata derived from the corresponding segments. It is a fascinating question how the nerves find the correct match. The growth cones of the nerve axons seem to follow a double-guidance system. Part is "global," in the form of substrate pathways that will be followed by foreign nerves if experimentally transplanted to the wrong segment. But in addition there must be a set of specific local cues that produce the final synaptic patterning. Recent work has concentrated on the possible roles of nerve CAMs in axon outgrowth and synaptogenesis (Tosney et al., 1986). An interesting paper by Stirling and Summerbell (1985) shows that if chick wing buds are experimentally reversed before axon invasion, in dorsoventral reverses axons will go to any target they meet passively, while in anteroposterior reverses the axons change path to find the "correct" target.

In vitro experimentation shows that, while isolated skeletal blastemata will differentiate to form an appropriately shaped cartilaginous element, finer-level details of morphology (defining characters at lower taxonomic levels) will be lacking (Thorogood, 1983). During the later stages of limb morphogenesis, interactive effects start to become important and perhaps the most significant of these in shaping the final phenotype must be interactions between the skeletal and muscle rudiments, leading to a functionally appropriate and taxonomically specific morphological configuration (Muller, 1986). Morphology does not become rigidly fixed with the completion of the ossification stage. There is extensive remodeling of the bony tissues of most skeletal elements throughout life (formation of trabecular systems, etc.) and even some changes in muscle relationships, all in response to mechanical conditions.

Although our data are incomplete, we can trace some general features of limb morphogenesis. We can see a series of interplays: between global and local controls, between intrinsic programming and extrinsic controls, between independent and interactive processes. We can trace out a series of key events: patterning of the ECM, epitheliomesenchymal interactions mediated by the ECM, reciprocal effects of mesenchyme cells on the ECM, setting up of patterns of cell density and behavior, physical constraints on the pattern of growth of blastemata, interactions among blastemata. In the next chapter we will use this and other examples to try to draw some generalizations about the properties of morphogenetic systems and their potential for change.

Evolutionary changes in the tetrapod limb are difficult to analyze on the basis of comparisons of adult phenotypes. We may take as a simple example the loss of carpal element. There is a very large number of ways in which this could be caused. Among these are failure of the prechondrogenic blastema to form, failure of prechondrogenic branching or segmentation, failure of a blastema to chondrify, or failure of a chondrified blastema to ossify. Failure of the prechondrogenic blastema to appear may be due to problems with pattern control, to problems of competitive recruitment to blastemata from a restricted mesenchyme pool, or to failure of cell division within the blastema. Perhaps, instead, the blastema does appear but then fuses with another. But if two blastemata are "fused" right from the beginning (assuming one can be sure that it is fusion), this could be due to specifics of the pattern control mechanism or to local effects (defects in the shape of the limb bud, for example).

It is, of course, very limited to deal with simple models of the limb skeleton that only model the patterns of condensation of blastemata. Whether or not blastemata stay separate or fuse may turn out to have a lot to do with the pattern of vascularization of the limb bud. If there is a blood vessel between two blastemata, they will tend to stay separate. This again may be controlled by whole-limb field phenomena, and again emphasizes that it is the whole pattern control mechanism that is "real" and fundamental, not its first-order epiphenomena such as blastemata.

The appearance of a wholly new element in the limb is a far less common evolutionary occurrence than loss of an element. As discussed further in the following chapters, additive change in the basic arrangement of the limb can only be accomplished at those very early morphogenetic stages where the mesenchyme becomes patterned. A particularly interesting set of limb variations are those of hyperdactyly and hyperphalangy where whole sets of elements are duplicated, or oligosyndactyly where there is reduction and fusion. These variants are readily produced by Goodwin and Trainor's model. They are also conditions mimicked in certain avian and mammalian mutants, which offer a useful way to check many aspects of models of limb pattern generation.

As reviewed by Hinchliffe and Johnson (1980), mutant conditions in the skeleton can be correlated with all of the "key" phenomena of limb bud morphogenesis listed above. They seem to be the result, particularly, of changes in the relative size and shape of the limb bud, redistribution of mesenchyme from one region of the bud to another with consequent excess and deficiency, and change in growth rates of blastemata. The study of mutants is therefore particularly useful in attempts to "dissect out" where control of morphogenesis lies. It is made complicated by the fact that a given phenotype may be produced by more than one cause, particularly in the case of fusions and regressions. Mutant conditions include eudiplopodia in chicks, where a duplicated limb is caused by a second parallel AER. This is a mutant acting at a very early stage. A whole range of mutants including the luxate mutants in mouse affects the size and shape of the limb bud (see discussion in Chapter 7) and thus the relative proportions of the limb and polydactyly and oligosyndactyly. Next there is a group of mutants that affects the proportions of the limb through differential allocation of mesenchyme to the different regions; brachypod in mouse, for example. Once the skeletal blastemata have been laid out, there are mutants that affect the relative growth of the cartilages. In nanomely in chicks there is a defect in ECM production. In chondrodysplasia in mice or dogs, there is a defect in collagen production. In creeper in chicks different skeletal regions show differential growth rates of cartilaginous rudiments.

While it is useful to concentrate on evolutionary changes in the limb skeleton, it should not be forgotten that change in any part of the limb must be accompanied by changes in the rest: bones, muscles, connective tissues, ligaments, blood vessels, nerves, and so on. The result of any change must be such as not to disturb seriously the functional integration of the limb. All changes must be functionally congruent. Perhaps the example of the evolution of major configurations of the reptile ankle will serve as a good example here. Paleontologists have long had problems with this example. It is well known that there are major different patterns of ankle joint architecture in fossil and recent reptiles. In the ancestral condition, the ankle joint is formed between the tibia and fibula proximally and the astragalus and calcaneum distally. In crocodilians (Figure 7), the astragalus has become attached to the tibia and the ankle joint

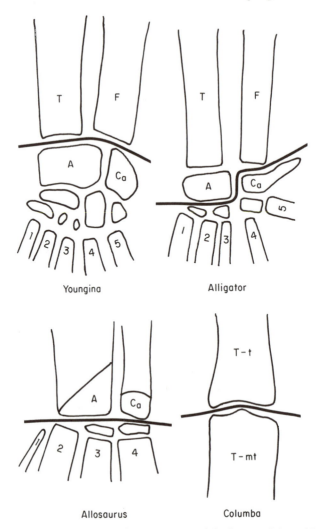

Figure 7   Four different patterns of arrangement of the bones of the ankle joint show-ing the different positions of the astragalus and calcaneum with respect to the functional joint. The line of joint flexure is marked by a heavy line diagrammatic. (A) The prim-itive condition, as shown in the Pernian eosuchian *Youngina*; (B) an oblique joint in the modern *Alligator*; (C) an advanced joint in the Jurassic dinosaur *Allosaurus*; (D) a more advanced condition in the dove *Columba*, with massive fusion of elements. A = astragalus; Ca = calcaneum; T = tibia; F = fibia; T-t = tibio-tarsus; (equals tivia + astragalus = calcaneum) T-mt = tarso-metatarsus.

now runs obliquely. In dinosaurs and birds, both astragalus and calcaneum are proximal to the ankle joint, which is now "intertarsal." One can see obvious functional correlates of these shifts in terms of limb mechanics and improve-ments in the gait. One can guess that such shifts would be under strong selec-tive control. But it is not possible that these changes could have occurred through a graded series of intermediates; the states are binary. Nor is it simply a matter of the skeleton. The muscles operating across the joint must shift their attachments. If the change were caused by an error in the pattern of the bones alone, the muscles of the limb, if they still attached to the "right" bones, would produce an ankle that could not flex. Only if the initial error either occurred in or forced a realigning of muscle blastemata with skeletal blastemata could a new ankle joint be created. The interesting feature of such a binary developmental shift is that the new phenotype could have been introduced in a single generation; the time taken to fix such a pattern in a population would then depend on sorting mechanisms, and could proceed either rapidly or slowly depending on the selective value of the whole phenotype in which the variant pattern occurred. Not only were the original variants fully functional, they turned out to be better than the old and formed part of the basis of new adaptive radiations.

A change in arrangement within a single functional unit of the limb such as the carpus and tarsus is perhaps even easier to imagine. Hanken (1983b) has produced a very interesting study of carpal and tarsal variation in a population of the terrestrial salamander *Plethodon cinereus*. In virtually all populations of this widespread species, carpal and tarsal patterns are highly constant. But if one compares different species within the family Plethodontidae, and even among related families of salamanders, one finds that they each have different characteristic tarsal and carpal patterns. Hanken discovered a Canadian population of the red-backed salamander, probably isolated from other populations and occupying a position at the periphery of the range of distribution of the species, in which tarsal and carpal variation is rampant. However, if the patterns of variation are analyzed, certain theoretically possible variations do not occur and others are in low frequency, while some are in high frequency. Certain of the latter are just those variants that characterize other salamander species. (Note that, given the finite number of variations possible, it is bound to be the case that variants in one group will match the characters of another.) Later (p. 89) we will discuss whether this example is a case of developmental or even phylogenetic constraint. Here we may examine what light it might shed upon the generation of variants under the strict pattern control of limb morphogenesis.

The first thing to note about the variants Hanken catalogues is that many of them are apparent fusions of blastemata (as no developmental data are available, one cannot tell at what stage this occurs). There seem to be very few outright losses, and one would guess therefore that the variants are not being produced in connection with change in limb bud shape and excesses or deficiencies of mesenchyme. It is probably also significant that the digital pattern is not variant. One possibility is that the variants are produced through errors in late morphogenesis forced by a speeding up of the developmental process due to a shorter "window" of suitable climate, or by a lengthening of developmental time due to cooler temperatures during development.

Perhaps one of the most important factors in production of variant phenotypes lies in the control of growth rates and the factors underlying them. In vivo and in vitro studies have demonstrated that different elements of the limb, at the condensation and chondrogenic stages as well as at the osseus stage, have different inherent growth rates. If these relative growth rates within the limb are altered, the whole pattern of the limb will be changed, while at the same time the appropriate muscle and innervation patterns can be maintained. If they are sufficiently changed, the musculature and innervation will be changed as well.

In such investigations a much-studied example has been the reduction of the fibula of the bird hind limb relative to the tibia, which retains normal "reptilian" proportions. In the presumed reptilian ancestors of birds, and in *Archaeopteryx,* the fibula is of "normal" proportions and its distal tip takes part in the ankle joint. In all modern birds, the fibula is reduced to a splint that ends far short of the tarsus. Hinchliffe and Johnson list several possible mechanisms to account for the change in relative size of the fibula and tibia: (1) differential recruitment to the blastemata at various stages from the neighboring mesenchyme, (2) differential rates of cell division within the blastemata, (3) different rates of production of intercellular matrix, (4) differential changes in cell volume in the cartilaginous stage, and (5) the absence of a distal epiphysis on the fibula, slowing down its growth rate. An often-cited set of experiments by Hampe (1959, 1960) seemed to support the first possibility—a competition model (see Wolff, 1958). Hampe inserted a mica plate between the tibia and fibula and in a small number of cases produced a reptilianlike (or putting it more dramatically, the *Archaeopteryx*) limb pattern with an elongate fibula (Figure 8). Hicks (in Hinchliffe and Johnson, 1980), however, found that the tibia and fibula had differential inherent growth rates and that in normal chick in culture a slow growth rate of the fibula is apparent even before epiphyses are formed. Archer, Horbruch, and Wolpert (1983) suggest that in Hampe's experiment, insertion of the mica plate prevented the fusion of the distal epiphysis of the fibula to the tibia and that normally it is this fusion that slows down the growth rate of the fibula. Most recently, Muller (1985, 1986; Muller and Wagner, 1987) has repeated Hampe's experiments and apparently solved the puzzle. Muller shows that the effect is produced by the fact that the fibula has a normal growth rate when a barrier is introduced, but the tibia has a slowed growth rate. But even more remarkably, Muller's experiments show that when the skeletal morphology of the limb is experimentally altered in this way, the muscle arrangements are also modified, and they also fall into a pattern mimicking that of reptiles. Three major muscle configurations had also shifted to a typically reptilian pattern. In other words, there was a morphologically and functionally coordinated set of changes to the phenotype in the formation of this "experimental atavism." Thus Muller confirms the interactive role of skeleton and muscles in modeling the phenotype of the limb.

THE VERTEBRATE HEAD

As a second example we may consider the development of the vertebrate head, obviously a far more complex structure than the vertebrate limb, consisting of multiple organ systems developing together. In all of comparative morphology,

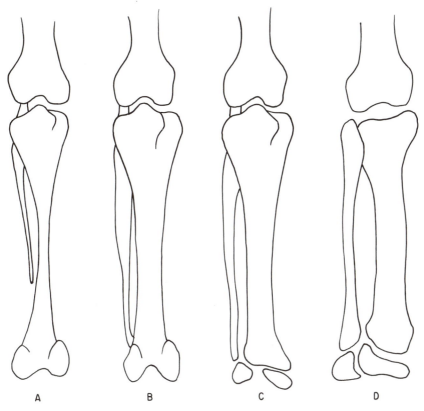

Figure 8 Comparison of (A) the normal bird lower hind limb with those of (B) an experimentally treated bird limb produced by Hampe and by Muller; (C) *Archaeopteryx*; and (D) a reptile (*Crocodylus*). Data in part from Muller (1985).

the vertebrate head has been studied more than any other system. Because it is complex it contains a great deal of information—a great number of apparently discrete characters that can be compared across taxa. Among fossil remains, cranial materials are relatively more common and therefore useful than other elements. But most importantly, the vertebrate head has been the focus of more theoretical studies than any other organic structure. This stems from the great and continuing debate over the possible segmental nature of the head, a debate started in the eighteenth century by Goethe and by Oken, who noted resemblances between the skull and a series of fused vertebrae.

That the vertebrate trunk is segmented is well known, although the segmentation is far less dramatic and complete than that of arthropods. But a segmental nature of the vertebrate head has never been satisfactorily established. The head consists of a number of important functional units: the brain and brain stem, the cranial sense organs, the cranial nerves, the neurocranium, the notochord, the vascular system, the apparatus of branchial pouches and clefts and associated visceral skeleton, a dermal skeleton and dentition, and the musculature of the head including that of the eyes and the whole facial-branchial apparatus. All these major features (and more) combine to make a structure that is complex and, among taxa, diverse in structure. They were acquired neither at the same time nor necessarily in response to the same pattern-generating mechanism. To unravel the principles and mechanisms of pattern formation in the head is a major and so far incomplete undertaking but one that is of considerable importance, for it is in structures of the head that so many taxa vary interestingly one from another.

In order to simplify the complexities of head structure, it is useful to go back and describe its unfolding in development. At the time of neurulation the vertebrate embryo has a fairly simple structure. As the neural tube forms, a new type of embryonic tissue makes its first overt appearance: the neural crest. In fishes and amphibians the crest is formed by an inductive relationship between the ectoderm proper and neurectoderm in the angle where the latter folds over. The crest is induced in the inner angle of the fold. In higher vertebrates, neural crest may in fact divide off from the neural tube itself. The neural crest will play a major role in the patterning of the head. Arising at the same time and through the same mechanism is the ectodermal placode system, which in tetrapods helps form the three paired cranial sense organs, and in fishes also forms a complex superficial laterosensory system.

Now there comes a cascade of morphogenetic interactions. The chordamesoderm, under reciprocal induction from the neural tube, differentiates into a notochordal strip in the midline with paired strips of paraxial mesoderm alongside. The mesodermal material that lies in front of the tip of the notochord, which means that it is the first part of the chordamesoderm to have passed over the dorsal lip of the blastopore, forms the prechordal plate. Subsequently the cells of this prechordal plate mesoderm form part of the posterior braincase floor (the parachordal cartilages). The paraxial mesoderm in the head is continuous with the paraxial mesoderm of the trunk. The latter will form the segmentally arranged somites. In the head region separate somites do not form. Instead there is a partial differentiation of the paraxial mesoderm into whorled cell masses, called somitomeres, that form a continuous strip of mesoderm on either side of the notochord and alongside the brain. Immediately above them, under the ectoderm, lies the premigratory neural crest. There is a variable number (perhaps four in anamniotes and seven in amniotes) of somitomeres in the head region, in front of the first discrete somite block of mesoderm. The first organization of the somitomeres in distinguishable in chicks before migration of Hensen's node, after which the somitomeres differentiate. The organization of the somitomeres is thought to be causally connected with patterning of the so-called neuromeres ("segments") of the early brain stem and reflects the basic anteroposterior regionalization of the whole body plan.

The branchial apparatus forms as a series of paired outpocketings from the foregut. It is not known under what influence these pouches form. If they are induced by the somitomeres, then the segmental appearance of the branchial pouches would be homologous with the segmental arrangement of the mesoderm and would be evidence of a single patterning process of the body axis.

There is no direct evidence at present (Thomson, 1988). The branchial pouches reach out toward the outer ectoderm and variously perforate—depending on the taxon concerned. As is well known, even in mammals, transitory gill slits form in this way (four pairs only). The gill arch skeleton forms in the spaces between adjacent pouches and ectoderm. Its mesenchyme comes from migrating neural crest cells, as does that of so many crucial elements of the head in vertebrates (see further, below). The theory states that each pair of gills has an arch skeleton behind it. Each gill unit also has a large sensory nerve from the brain (sensory cranial nerves V1, V23, VII, and IX–XII). Each gill has a muscular component for ventilation of the gill chambers and these muscles survive as the facial, tongue, and pharyngeal muscles in mammals. As is well known, the jaws of gnathostome vertebrates form as modifications of the skeleton of the original first gill arch, and therefore the jaw opening and closing muscles are part of the basic branchial series.

The other major muscles of the head are the three pairs of eye muscles. These are formed from anterior cranial somitomeric mesoderm, with possible contribution from prechordal mesoderm. The jaw and branchial muscles are formed from successive cranial somitomeres and the first true somite blocks.

As the somitomeres start to differentiate, the neural tube undergoes its own regional differentiation, especially under the influence of the three sets of paired cranial sense organs. The forebrain develops in association with the nasal placodes, the midbrain in association with the optic placodes, and the hindbrain with the otic placodes.

The skeleton of the head is divisible into three principal units. We have already mentioned the gill arch skeleton. The other two units are the neurocranium (a series of ossifications and cartilages surrounding the brain and notochord) and the dermal skeleton. The dermal skeleton is formed directly in the dermis without cartilaginous precursors. It essentially covers all the outer surfaces of the head. This means that not only the external surface of the head has a skeleton of dermal bones, but all the lining of the mouth as well, the inside of the lower jaws, the underside of the primary upper jaws, the floor of the braincase, and the innner surfaces of the gill elements. It is on those bones that line the mouth cavity that the dermal teeth are formed. In primitive vertebrates the dermal skeleton is structurally complex, with outer enamellike and dentinelike tissues that are formed from a complex interaction between mesoderm and superficial ectoderm. The specialized teeth of more advanced vertebrates have evolved from this early complex dermal skeleton.

The neural crest is a sort of invisible glue holding all the skull together in a figurative if not literal sense. The neural crest arises as a discrete embryonic tissue and then starts to disperse as its cells migrate along largely predetermined pathways to invade every part of the head. A comparable trunk crest exists in every segment of the trunk. The cranial neural crest either forms or contributes to nearly every feature of the head. It forms the ganglia of the sensory nerves, the connective tissue sheaths of all the craniofacial musculature, pigment cells, the odontoblasts of the teeth, and a great deal of the cranial skeleton. The neural crest forms the whole of the visceral skeleton—that is to say the gill arches, with the curious exception of the second basibranchial bone in amphibians. It forms the ventral and anterior part of the dermal skeleton and also a small portion of the neurocranium. By contrast, the trunk neural crest, while also contributing to sensory ganglia, pigment cells, and gland cells, apparently has almost no skeletogenic function, at least in tetrapods. Trunk crest is reported to form the connective tissue of the dorsal fin in amphibian tadpoles and the median fin skeleton of lampreys, but nothing is known of its possible role in the formation of the dermal head or trunk skeleton of fishes.

The path of migration of the neural crest cells is obviously crucial to the morphogenesis of the vertebrate head. Anderson and Meier (1981) show that the initial spatial arrangement of the cranial neural crest, and its early migration, are in part controlled by the configuration of the somitomeres over which it lies, and by the shape of the brain and sensory capsules. As development proceeds, the neural crest migrates in two general pathways in the head (three in the trunk): outward between the ectoderm and the somitomeres, and downward between the inner faces of the somitomeres and the neural tube. In both paths, as already mentioned, it migrates in defined cell-free spaces rich in ECM. The migration of crest is heavily influenced by the architecture of the rest of the developing head—particularly the brain, somitomeres, and sensory capsules—around which it must find its way. Noden (1986) has detailed a case where a defect in the spatial patterning of neural crest from the rostral plate area causes major facial malformations in a highly selected breed of cats.

Neural crest disperses to the dermis first and thus forms the most anterior and distal parts of the dermal skeleton; it is followed by dermatomal material from the somitomeres, which presumably migrates along the same general pathways under the ectoderm (Thomson, 1987b).

The somitomeres of the head evidently function pretty much like the trunk somites in having dermatomic (connective tissue and dermal bone of the dermis), sclerotomic (skeletal material of the endoskeleton), and myotomic (muscle-forming) regions. Growth of somitomeric mesenchyme to its target organs seems to follow after neural crest migration has started. The migration of mesodermal mesenchymal material has not been as closely studied as that of the neural crest ectomesenchyme, but presumably follows a similar pattern.

The joint formation of the craniofacial musculature from a mesodermal core and a connective tissue sheath derived from neural crest is extremely interesting. Lateral plate mesoderm is missing from the head, and in a sense the neural crest substitutes for it in the head. In fact, neural crest forms virtually all the connective tissue of the vertebrate head.

Early Evolution in the Head

The origin of vertebrates from some unknown chordate ancestor (taking the premise that amphioxus is the primitive outgroup) was accomplished by the evolution of several major innovations: (1) the neural crest (absent in amphioxus and all protochordate relatives); (2) bone, cartilage, and other hard tissues (in connection with the neural crest); (3) the three major cranial sense organs, nose eye and ear, and related placodal systems; (4) the forebrain; (5)

the embryonic prechordal plate); (6) a single ventral heart and associated vascular pattern; (7) glomerular segmental kidneys. Hypothetically, the origin of vertebrates was connected with an increase in size and level of metabolic activity in an active free-swimming form living in shallow oceans, perhaps on or near the bottom where there was a supply of detrital food material. This ancestor used its branchial apparatus of pharyngeal pouches as a feeding rather than respiratory organ (Thomson, 1971; Mallatt, 1984; Northcutt and Gans, 1983).

A hypothetical vertebrate ancestral stage, being larger and active, evolved a large brain in connection with the evolution of the three paired cranial sense organs, which presumably had their precursors in chemosensory, light-sensory, or statocystlike structures, already passing sensory information to the brain stem. Perhaps the first innovation was the enlargement of the forebrain region and a change of inductive signaling such that the first elements of mesoderm to pass over the dorsal lip of the blastopore at the front of the presumptive notochord came to have the role of inducing the formation of the forebrain structures. This then evolved into the prechordal plate. Perhaps next, after the organization of the eye and ear, came evolution of the neural crest, starting perhaps from a precursor in a peripheral sensory nerve net (as suggested by Northcutt and Gans, 1983). The neural crest may first have contributed to the sensory nerve ganglia associated with an enlarged branchial feeding apparatus. How it acquired a general mesenchymal role and particularly a skeletogenic role is a total mystery.

If von Baer's laws are useful, then we might suppose that there was an early stage in vertebrate evolution when the only skeleton present was that formed by the neural crest and represented today by present distribution of ectomesenchymal skeletal tissues. The problem of the dentition and the relationship of the neural crest to the odontodes in the head and trunk placoid scales and other dermal features complicates the story greatly. Perhaps the odontogenic function came second and the capacity of the mesodermal mesenchyme to form cartilage and bone third. However, once the genetic information to produce hard tissues was within the genome, it could potentially be expressed anywhere and this would suggest that ectomesenchymal and mesodermal mesenchymal hard tissues appeared at essentially the same time, giving us no idea as to which appeared first or where. Perhaps we should instead think first in terms of connective tissue and only secondarily in terms of hard tissues. In this case, we might guess that the neural crest provided the connective tissues to the head and the mesoderm provided it in the trunk. Connective tissue would have been the first "skeletal" material, made necessary by increase in size, apart from the notochord itself, which is of course a muscle in amphioxus and in larval lampreys (Guthrie, 1958).

Pattern in the Vertebrate Head

Pattern formation in vertebrate head morphogenesis appears to involve a mixture of global and local factors, just as in the case of the tetrapod limb. Many of the controlling factors are evidently set in place early in development, even at the blastula stage and at gastrulation itself: for example, cell numbers and growth and division rates. There are three major morphogenetic elements.

1. Ordering in the anteroposterior axis occurs as part of the general antero-posterior ordering that is obvious first at gastrulation, but evidently starts well before because both the neurectoderm and the paraxial mesoderm (and perhaps even the neural crest region) are ordered before gastrulation occurs. By the time gastrulation is finished and the mesoderm and neural crest start to differentiate, the whole head has been laid out in an anteroposterior axis and most morphogenetic events follow an anteroposterior sequence.
2. Lateral, dorsal, and ventral migration of the proximal mesenchymes seems to be directed by gross mechanical configurations in the cell-free spaces, formed between the major organ rudiments, by the relevant ECMs, and by a general drive from high to low cell density (i.e., to the periphery).
3. Differentiation of the brain, the paraxial mesoderm, and the branchial apparatus imposes a gross order on the whole system, as do the three cranial sense organs. There is at least the semblance of a segmental patterning in the vertebrate head; neuromerism, metamerism, and branchiomerism may in the end be shown to reflect a common causal mechanism. But even if they are not, there is a convergence in their sequential patterning.

Analysis of the detailed control of morphogenesis in the head is far behind what is available for the tetrapod limb, in large part because of the immense complexity involved. However, we can be sure that there is much general similarity between the two. An early factor is the laying down of ECMs, especially as they will later both control the pathways of cell migration and mediate epitheliomesenchymal interactions. The patterning of mesenchyme outgrowth and migration will also be controlled by cellular and tissue-level architecture in the developing head complex. There is much interaction between developing units, especially of an inductive sort. Most organs derive their tissues from multiple sources. (For example, the connective tissue of facial muscles comes from neural crest ectomesenchyme and the myoblasts from somitomeric mesoderm.) There must be a complex interplay between global and local controls, and intrinsic and extrinsic controls.

The question of segmental or other types of patterning in the head is important because it concerns the base of much of what we use to recognize homology in evolution. Not only do we trace homology by finding transformations in series of real structures (the stapes in reptiles and mammals, for example), but we also make reference to an underlying segmental theory to argue the case (e.g., that the stapes in reptiles and in mammals represents the epibranchial element of the first visceral arch common to all vertebrates).

Here an experiment of Noden's may be instructive. In general, neural crest in the vertebrate head has been thought to be pluripotent. That is to say, the crest cells are capable of forming a wide range of different structures. This pluripotency has been thought to extend beyond the stage of the initial migration. What they finally form has been thought largely to depend on the local environments into which they migrate and within which they differentiate. Obviously this is an oversimplification. Cells will not migrate even roughly in a particular direction without some internal controlling mechanism. It may merely mean that in most experimental manipulations, the initial specification of migrating cells can be overridden by an abnormal environment. It does not mean there is no internal prior specification of populations of migrating cells. Noden (1983, cf. 1984) performed an experiment in which the neural crest that would normally form the first skeletal arch (i.e., the jaws) in chicks was transposed to the position of the second arch (that of the hyoid apparatus and the columella of the stapes). The expected result was that the heterotopic crest population would be transformed and form a typical second arch system. Instead it formed a duplicate set of first-arch structures, and even an ectopic beak. In other words, here the specification of a population of cells from a particular region of the anterior neural crest was present before migration and was retained in a foreign environment, even down to forming quite accurately shaped elements (squamosal, quadrate, Meckel's cartilage) in the wrong place. The properties and whole configuration of the hyoid arch environment were not sufficient to regulate the differentiation of the heterotopic transplant to the hyoid pattern.

This is a quite remarkable result in that one would not expect migrating crest cells, passing individually from the proximal neural tube, to have sufficient information alone to control their own precise spatial ordering in masses of mesenchyme condensations and the shaping of resultant skeletal structures.

The results of the experiment contain some anomalies. For example, a second external auditory meatus is formed. Normally the external auditory meatus is thought to represent part of the hyoidean gill opening. The regular opening forms, but with a second blind external structure in front of it. On the other hand, the columella of the stapes (a second-arch structure) is not formed. Further, Noden (1986) has found that neural crest that would normally be destined for the frontonasal process of the dermal skeleton will also form ectopic first-arch structures when transplanted to the second-arch region. The great importance of the experiments is to draw attention to the question of intrinsic versus extrinsic control in morphogenesis (see also Kay, 1986).

Growth and Form

Throughout development, presumptive tissues and organs have two sets of properties, often changing and always under precise control. These are (1) the state of commitment toward a particular mode of cytodifferentiation and (2) particular rates of division and growth. As morphogenesis proceeds and the state of commitment of cell types comes closer to final differentiation, aspects of relative growth become more and more important in literally shaping the final phenotype. In both the vertebrate limb and the head, the number of cell, tissue, and organ types is specified relatively early, but relative proportions (size and shape) continue to change througout morphogenesis until growth stops (not always with sexual maturation), according to fixed allometric relationships (Huxley, 1932). It is here that an enormous potential for production of evolutionary variants is seen, as d'Arcy Thompson first showed. de Beer and Gould showed the power of heterochronic fixation of such changes. Study of allometry and heterochrony has become a major element in macroevolution in recent years.

Some changes in relative proportions are shown to produce only morphologically localized and taxonomically restricted innovations. For example, allometric changes lie behind the relative minor shifts causing the difference in skull shape between chimpanzees and humans (Gould, 1977), or among the species of deer (Gould, 1973). Other changes have more far-reaching consequences. For example, we will see below that changes in the relative size of the cranial vault and in the proportions of the jaws lead to major changes among mammallike reptiles, leading to the origin of mammals and, among other things, the origin of the mammalian middle-ear ossicles. What is lacking from such studies is knowledge of how allometric relationships are controlled. While heterochrony allows the fixation of significant phenotypic differences based on common allometric patterns, a much more interesting cause of phenotypic change exists in the reprogramming of allometry itself, rather than in the working out of a given relationship over a range of sizes.

7

Some General Properties
of Morphogenetic Systems

J. Maynard Smith (1983) has written that "although we have a clear and highly articulated theory of evolution, we have no comparable theory of development." I would turn this statement around somewhat and say that until we have a general theory of development we are unlikely to be able to derive a complete theory of evolution. This does not mean that a theory of evolution is wholly contained, in some reductionist sense (see Chapter 2), within a theory of development. However, if developmental processes play a major role in determining the modes and tempi of introduction of new variation at the level of the individual organism, and if they also have roles in upward and downward causation to other focal levels in the hierarchy of evolutionary mechanisms, then at least some of the rules of variation must be contained within the rules of development. If we are to progress in evolutionary biology beyond the study only of the contingent, and of unique empirical events, we will need a general theory, and part of that theory must derive from theories of the developmental processes that drive the introduction of variation. From developmental theory we will be able to make new general statements about how variation can be introduced at the phenotypic level. Although we still lack any such general theory, we can begin the process by using the preceding discussions at least to propose some general properties of developmental systems.

The properties and processes of morphogenesis form an extremely complex system. Perhaps the hardest parts to grapple with are those "whole-organism" properties by which any given region of the developing embryo responds to field phenomena created by and expressed within the organism as a single whole rather than as a collection of isolated units, each with their own independent problems, mechanisms, and histories. Although these are vital questions (no pun intended), relatively little experimental work is being conducted in this area for obvious conceptual and technical reasons. Reductionist, functionalist approaches tend to make one concentrate on the parts rather than the whole. Holistic approaches, while interesting conceptually, are harder to use as the basis for practical experimentation on real systems, although there is a multitude of questions to ask.

On the other hand, research into the regulation of gene expression in developing systems has proceeded apace. The techniques for assaying gene activity and for analyzing genetic mechanisms that have been developed in the last 20 years have contributed enormously to our understanding of the action of a small number of genes within a relatively small number of taxa. For example, quite a large body of results has built up concerning the regulation of ribosomal RNAs such as 5S and 18S in amphibians. But even so there are enormous gaps in our knowledge, particularly in the area of the genetic control of the patterns of development. There are some interesting models of mechanism of gene activation, but we are a long way away from understanding the full temporal pattern of activation and regulation of the genome in development from first cleavage to final phenotypic expression. The "differential gene expression" hypothesis predicts that during the course of development from egg to embryo, within any cell lineage the genome is first repressed, then selectively derepressed and activated transcriptionally in a complex pattern; on top of this a second pattern of actual gene expression may be imposed through translational control. But we know very little of the overall pattern of control systems—how the system actually works, which means how it controls itself. We know even less about the mechanisms of morphogenesis at the tissue level that both result from and cause interactions among cells (expressing their current state of gene activity) and their environment (largely the activities of other cells).

In these circumstances, discussions of the control of development are bound to be more than a little premature and incomplete. But a surprising amount of data is available concerning bits of one puzzle here, another there. And we can begin to establish what some of the general patterns and processes might turn out to look like. In this chapter I will discuss some aspects of the dynamics of developing systems, their control, and therefore their potential for change.

CONTROL SYSTEMS IN MORPHOGENESIS AND CYTODIFFERENTIATION

The simplest and most useful way to envisage the control of developmental processes is as a cascade of "decision points" at which, using the terminology that we have discussed previously (Chapter 4), the fate of a cell lineage becomes specified and then more and more committed. Toward the end of this long chain of events, cell lineages actually become determined; that is, cytodifferentiation of the final phenotypic characteristics begins. At each decision point each cell and lineage encounter a "switch" and respond to one of two (or perhaps even more) possibilities the results of which (a) initiate new gene expression, (b) materially alter, often in extremely subtle ways, the behavior of the cell during morphogenesis, and (c) progressively restrict the potential range of future differentiation of that cell and its lineal descendants. The accumulation of genetic expression via the passing of these switch points constitutes a "memory" (see Wolpert's "positional information" concept, Chapter 4), "genetic address" (Stein, 1980), or as Slack (in Smith and Slack, 1983) puts it, a "second anatomy" of the organism.

These sequences of decision points give us a picture of a "developmental pathway" in that we can in theory trace backward from a finally differentiated tissue to reconstruct all the switching points that have been passed and that have contributed to differentiation. They form a trajectory through the complex of intersecting signals and responses that make up the whole of development. Consider, for example, a stream of migrating neural crest cells in the vertebrate head. They (or their ancestors in the cell lineage) have already been through a series of decisions specifying that they will constitute a mesectoderm, that they will have migratory properties, that they will respond to the presence of extracellular matrices in migration, and they will later be able to differentiate into a range of tissues including connective tissue, cartilage, and neural ganglia. In order to reach their final differentiated states, neural crest cells will need to experience a whole set of signals stemming both from global properties of the whole organism and from microscopically precise interactions with other tissues that are themselves undergoing progressive commitment and determination. All of these factors, and many more as well, contribute to forming the particular environment within which the neural crest cells of the head migrate and differentiate. And at the same time the crest cells are themselves adding a set of components to the system.

Genetics, Epigenetics, and Self-Assembly

Such a system of cascading geometric growth in complexity over time, has two important properties. The first is, of course, that the whole system is under genetic control. Genetically coded information starts the process going, and at every step in the way new information is expressed and other elements of gene expression are switched off. There is, at it were, a genetic script that the whole cast of characters is following, a script that is creating the cast of characters as the plot moves along. But at the same time, especially in strongly regulative developmental systems, any developmental process is also a product of the environment in which the play is acted out. This environment is, once again, in large part created as a result of the other information in the genetic script. The result is a sort of interplay between two phenomena, genetic and epigenetic. As a result of relatively discrete genetic information, particular gene products are expressed. These then form signaling points in the complex array of events happening within the embryo. What results from the input of this new information is then epigenetic, falling under a new whole range of controlling influences different from those processes that caused the initial gene expression. As Waddington put it, epigenetics is "the causal interactions between genes and their products which bring the phenotype into being" (1975).

To some extent all developing systems, particularly in strongly regulative systems, are in large part self-controlling. The products of gene expression do not fit neatly into place like the components of a car on an assembly line (here a steering wheel, there a brake drum), although strict mosaic development almost has this appearance. In regulative systems, complexity grows out of itself. This is probably the only way that a truly complex organism can be created—through an exponential geometry of information. At every moment in development new information is being added in the form of new gene expression, and each increase in information is immediately multiplied manyfold at the epigenetic level because of the way in which the new quantum of information changes the nexus of interacting signals. The amount of information in such complex and interactive self-assembling and self-regulating systems is far greater than the arithmetic sum of all the discrete "bits" of input. Even so, in such systems, many phenotypic features are not fully determined at the genetic level. A classic case is that of fingerprints. The fact that each human being has a unique fingerprint is well known. The amount of discrete genetic information that would be needed to code for a unique phenotype for each individual is impossibly large. Instead, what is inherited is a capacity to make fingerprints according to a number of potential combinatorial properties. The fingerprints then self-assemble during development and, because the potential number of combinations is so large, each is unique. The same is true of the detailed "wiring" of the human brain. Each potential neuronal connection among millions of neurons could not possibly be under unique genetic control. What is inherited is a general pattern out of which we each have self-assembled our own, and so to this extent our intelligence is partially self-assembled rather than inherited (Hutchinson, 1981).

Central to the whole system is a complex feedback between the genetic and epigenetic components of the developing system. Information is added to the system in the form of expression of particular gene arrays. These then modify the epigenetic environment. This environment then calls forth a new set of gene expressions, and so on. At each stage a new level of irreversible decisions has been made by the embryo in its steady progress toward final determination of all cell and tissue types. Such a system is essentially hierarchical, operating through a series of hierarchical stages, each dependent on the ones before. We can define a series of focal levels at which the stage of commitment of a given tissue is defined on its own terms (essentially as the state of gene expression within its cells). At each level the tissue exists within an environment partially of its own making in which certain new possibilities are created that were absent before (emergent properties) and could not be created by the tissue acting alone. By interaction with that environment, the tissue moves to a new state of commitment and the cycle begins again. At final differentiation, the tissue is presumably no longer capable of reacting to its environment by means of change in what has become a steady state of gene expression.

Some General Properties of Pattern Control Mechanisms

The essential feature of any new stage in a developmental pathway or sequence is that a new pattern of gene expression is caused. Change in gene expression in a cell lineage changes the behavior of each cell and its descendants (except as further modified again). The lineage is sent off in a subtly or radically new trajectory in which cell behavior is controlled partially by the nature of the cells themselves and partially by the epigenetic environment (global and local) in which they find themselves. Control in a development pathway is therefore exerted in two places: (1) in the causation of new gene expression, and (2) in the control of the subsequent function of the newly defined cell lineages. Control of the first type is likely to be in the form of a discrete signal, at least in the sense that it is short in duration and specific to the signaling and responding tissues. Control of the second type is likely to be longer lasting and nonspecific. As we have already noted, in the morphogenesis of highly regulative animal systems (e.g., vertebrates) the signal that triggers new gene expression is often nonspecific in the sense that it does not work through the synthesis of some specialized signal molecule created solely for the purposes of giving a particular signal at a particular time to a particular target. Instead, the target system is likely to respond to a more generalized signal that represents an intrinsic property of the signaling source. The specificity of the signal–response system therefore comes as a property of the responding system rather than the signaling system. A given "signal" may trigger different responses in different systems at the same time; this would be true of many responses to gradients, for example, including the AER and ZPA gradients of the tetrapod limb bud.

Whether or not gene expression is triggered by specific signals, the following period of cell behavior is controlled by a whole range of new signals. Some will be relatively specific, contributing to further stages of commitment. Others will be quite general, for example in controlling rate of cell division or migration. Part of the result of the new gene expression must indirectly help create the environment in which the same cell lineage develops. All will in some way or another be signals to which other cell populations will respond. Each new set of gene expressions contributes to the overall morphogenetic patterning of the embryo. It cannot occur in isolation, and the results are not effected in isolation.

The cause of pattern in the embryo will have three main components: It will be (1) controlled by a series of developmental rules appropriate to the given focal level in the developmental cascade, (2) controlled by the boundary conditions within which these rules operate for the given set of cell populations, and (3) affected by the initial conditions relevant to that particular cell population in a given time and space (Figure 9).

Developmental rules, as we have already briefly noted (Chapter 3), are of many sorts. Among the most powerful are structural rules. A favorite example is the fact that the patterning of leaf primordia in a growing plant shoot or flower often follow the classic Fibonacci series (Kauffman, 1983). The growth of the shell in molluscs follows certain set mathematical relationships, and differences in shell shape among taxa can be related to quite simple changes in these relationships (Thompson, 1917; Raup, 1966). The later stages of morphogenesis in most organisms show fixed allometric relationships between the parts. Structural rules range from the most elementary (the cube–square relationship of linear to volumetric measures) to extremely complex combinatorial codings. Other developmental rules are the rules of materials, constants such as diffusion constants, the gas laws, or even the gravitational constant. These developmental rules set a developmental context that both defines and confines the limits of developmental processes and therefore of phenotypic results.

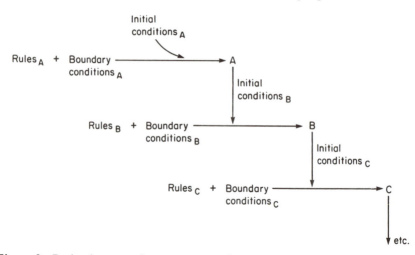

Figure 9 Basic elements of a pattern control system at two consecutive levels in a developmental pathway.

The developmental rules are, as it were, the "rules of the game." But how the game is actually played depends also on the players or, in this case, the boundary and initial conditions. In any developmental process such as the folding of the neural plate, the developmental rules (here rules of folding sheets, rules of cell division in a single layer, rules of cell-shape change in a single layer) are expressed in a real situation that delimits the boundary conditions. In the case of the neural plate, the boundary conditions are those of neurectodermal cell properties, or relationships to particular ECMs and neural CAMs of the neural epithelium, and the nature of the inductive signal that triggers folding. In part they are the properties that would be likely to be expressed in common by any neurectoderm (even those grown in culture). The initial conditions, however, also have a major role in shaping what actually happens.

The initial conditions are those that are unique to the particular event. They are, for example, the initial cell number or density, the patterns of growth and shape change of neighboring structures or, in the case of the vertebrate neurula, the whole embryo. They include the nature of the inductive signal given by the chordamesoderm, its strength and signal time, and the local details of ECM quantity, quality, and pattern.

At each point in a developmental cascade at which new gene expression is caused, and through the following stage of cell behavior until a new array of gene expression is caused, developmental rules, boundary conditions, and initial conditions will combine in the same way, as shown in Figure 9. At each stage they set a manifestation of pattern and process that then forms the new boundary conditions for the next step in the cascade, where new rules and new initial conditions appropriate to that level will also operate. These new rules and initial conditions are themselves predicated in part by all earlier events.

No developmental pathway operates independently. Rather, the boundary and especially the initial conditions in any level are likely to be strongly affected by what is happening in other pathways. In the nexus of interacting influences out of which the whole array of developmental pathways is carved, boundary conditions and the rules applying to those conditions are largely determined within a given pathway, but interactions with other pathways will potentially have a major effect on the initial conditions. The other major role component of interaction between developmental pathways must be in providing crucial signaling at the switch points. As we have seen, some of the signals that trigger the initiation of a new level of process are inherent to that unit and there is a feedback of signal, cause, and effect that directs morphogenesis. An example of this might be the relationship between cell behavior and extracellular matrices in limb development. Other signals are wholly extrinsic, for example the signal between heart and eye in the development of the latter. The signaling system is therefore strongly historically contingent and opportunistic, as well as ranging from strongly instructive to weakly permissive (below).

Thus we can build up a picture of a developmental cascade creating a pattern progressively through the operation of a network of controlling principles and signaling systems.

In such a system, if there is change at any point, the consequences must be felt at all points further along the cascade unless they can be specifically buffered or removed. If there is an increase of information at any stage (for example, a signal might produce two different responses within a large cell population instead of the expected one), the effect must potentially be multiplied as the cascade progresses (Figures 10 and 11).

In a complex cascade there is a lot of room for what we might call "reprogramming." That is, the system of signals and responses can be refined and modified in evolution either to produce a new result or, indeed, to sustain the

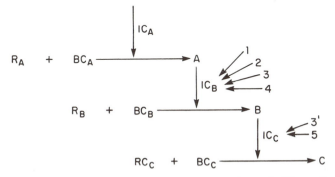

Figure 10 Modification of the pattern control system shown in Figure 9, yielding a divergence in pathway B to produce both B and B. These then interact separately at the subsequent stage in the pathway producing C (normal) and a new stage along the divergent pathway. R_A = Rules of A; BC_A = boundary conditions of A; IC_A = initial conditions of A; similar notation for B, C, , and . R_A = rules of A; BC = boundary conditions of A; IC = initial conditions of A; similar notation for B and C. Numbers 1–5 represent influences from five different intersecting developmental pathways.

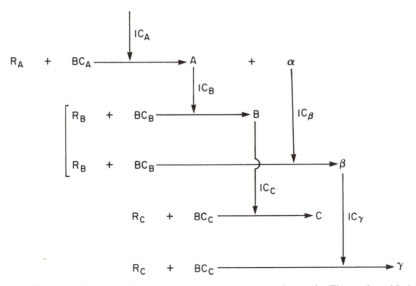

Figure 11 Amplification of the pattern control system shown in Figure 9, with interactions between the first pathway and neighboring pathways, showing the points of possible interaction.

original one. We can see that this must have happened in the evolution of Amphibia. In the induction sequences of eye lens in different taxa of anuran amphibians there are different numbers of stages of necessary signal. What is instructive in one taxon may be merely permissive in another. What is induced in one taxon may autodifferentiate in another. We see here an array of different ways to attain the same objective. Similar reprogrammings, through addition, deletion, or substitution of signals in the cascade, must accompany all evolutionary changes (we will discuss this further in Chapter 9). The result may be that a developmental pathway in a given organism is hardly optimized and refined (parsimonious). Instead it may be extremely arbitrary and sloppy, almost a random walk from one fixed reference point in the decision cascade to another.

The Signaling System

Although in theory we can assemble a list of all the interactions that form the "developmental pathway" for a particular tissue, in another sense, to do so is almost impossible. This is because in a highly regulative system such as vertebrate development, the list of potential contributing factors would be no less than a list of everything that ever takes place in the whole embryo, and one should probably be wary of notions like "housekeeping" and "developmental" genes except as an index of relative position in the developmental hierarchy (see below). What is "housekeeping" at one level may earlier have been a crucial step. In simpler patterns of development the course of events is perhaps more definable, but while is is one thing to delimit all the decisions that seem to be being made, it is quite another thing to understand what the signals and responses are: what the developmental control mechanisms are and how they work.

During the normal course of development, each decision point has a cascade of consequences. Primarily it is the direct source of control over gene expression transcriptionally or translationally. Each decision also delimits the range of future decisions that will be available to the cell lineage concerned. This limitation has twin aspects: (1) proscriptive, closing off certain ranges of future decisions, and (2) predicative, opening up a particular range of future decisions.

As the complexity of the embryonic ground plan builds up in ontogeny, the range of interactive signaling that a presumptive tissue region will encounter is more and more restricted. For example, timing and topography make it almost impossible for neural crest cells in the vertebrate head to be subject to the set of signals that activate the gene sequences that produce the differentiation of notochord from chordamesoderm. If they did, they could not respond anyway, because they have previously been committed to a different course. Once neural crest cells have started to migrate, the range of signals they can encounter is limited. The gene arrays necessary for differentiation into cartilage precursor cells are not switched on in trunk crest cells even when they are experimentally transplanted alongside cranial crest cells and are migrating normally with them. Is this because they still are not receiving the normal "on" signal (i.e., the experiment is faulty), or because they potentially could be switched on (perhaps were in some ancestral group) but would need a different "on" signal that is not expressed (either in head or trunk)? Or is it because the set of decisions that originally defined the trunk crest included proscriptive effects blocking future activation of those gene sequences?

Another way of expressing the apparent proscriptive aspect of decision points is that most decisions are irreversible. This again reflects the fact that the pattern of decisions is hierarchical. Decisions of equal hierarchical level might be interchangeable, but decisions taken to new hierarchical levels are not.

Probably rather few of the control signals in development are likely to be completely extrinsic to the embryo. However, there will be a range of quite crucial early decisions in development that are controlled by factors external to the oocyte and affecting the localization of cytoplasmic determinants in the egg. There may be other important maternal effects in terms of external gradients to which eggs and embryos are exposed if retained within the mother's uterus. Smith (1985) shows, for example, that embryonic axis orientation in the mouse is correlated with the orientation of the blastocyst within the uterine horn. The timing of events in larval development and metamorphosis may be externally mediated. There is also an extensive literature showing that experimentally induced environmental shocks—heat shocks, for example—can materially affect development. But the extent to which direct environmental factors affect normal development is little understood. Study of the interaction between ecology and development remains a whole subject to be developed in the future for, as van Valen (1976) has expressed it, evolution can be defined as "the control of development by ecology."

The vast majority of the signals to which cells respond at decision events must be intrinsic. In a wholly determinative system cells do not have a choice of responses but rather follow a sequence of decision points. If cell division has correctly apportioned cytoplasmic determining factors, cells will not be faced with conflicting signals. In regulative systems a given cell might be subject to more than one signal if, for example, it is right at the margin of two presumptive tissue types responding to separate signaling influences. A given cell may also potentially be subject to a range of signal strengths, and it is likely that the response of the cell could vary according to that signal strength. To take the example of the neural crest once again, the crest and the epidermal placodes are both formed by induction between the neurectoderm and ectoderm; the difference between the two may be caused by the spatial relationships of different subregions to the same inductive signal of which they receive different strengths.

In regulative development each set of decisions that is taken creates a new set of signals and potentially responding conditions. No signal can be correctly perceived unless particular decisions have already been taken by the receiving cells types. The whole system forms a sort of spiraling nexus of signal-response-signal. But every system needs a minimum of information to get it started, and this probably always comes initially from a mosaic of determinants inherited maternally. At the minimum these set up differentials between adjacent cells and across the whole embryo, and these then form the basis of a geometrically increasing spread of complexity in self-signaling.

We do not yet have the requisite data to explain the pattern of genetic control of complex morphogenetic systems, but there have been interesting conjectures concerning the coding of the switches involved in sequential pathways of development. If we assume that each decision point represents a binary switch at which a gene or set of genes becomes either in the "on" or "off" condition, we can build up different patterns of codings to account for the patterns of development. A limited but nonetheless useful view of the problem is given by the switch mechanisms involved in the morphogenesis of segmental structure in *Drosophila* via the famous homeotic mutants in the bithorax complex. Here we have the special case of the serial differentiation of a single territory into ten subterritories (the two posterior thoracic segments and the eight abdominal segments). Lewis (1978) explains the genetic control of these by postulating a pattern of serial coding. In the mesothorax all the bithorax complex genes are in the off position; in the last abdominal segment all nine are in the "on" condition. In the intervening segments the conditions are strictly serially, additively, graded. In such a scheme all nine switches could be responding to thresholds in a single anteroposterior signal gradient. Kauffman, Shmyko, and Trabert (1978) propose that the same conditions could be accounted for by fewer integrator genes if they were activated according to a combinatorial code. Whereas in a serial coding, n different conditions requires $n - 1$ switches, in a combinatorial system, n binary switches will code for $2n - 1$ conditions. In this case four switches would be needed and actually there would be five conditions spare (2 4 − 1 = 15). However, as Slack (1983) points out, the apparent economy of combinatorial coding is offset by the fact that a combinatorial code cannot respond to a single gradient and in fact for n conditions one needs n different signals to throw the switches.

COMPLEXITY AND STABILITY

Development is quintessentially a normative proceeding; its goal is to produce the "standard authorized edition" with as few misprints as possible. At first sight this seems incompatible with its complexity, especially the complexity of interactions in regulative development. One could view the great complexity of developmental cascades as being like the complexity of a sophisticated electronic system. The slightest particle of dust should cause it to malfunction. The difference is that development is in large part a self-assembling process and increasing complexity may in fact, almost paradoxically, lead to greater capacity to deal with errors and malfunctions. Not only is regulative development essential for the development of any great degree of structural complexity, by virtue of its accumulated blind complexity and interactiveness, it is essentially buffered and self-correcting against error. If so, then it is in such properties that we can discover a major capacity for evolutionary change, as we will discuss in this chapter and again in Chapter 9.

Canalization

In his many writings on development (e.g., 1975) C.H. Waddington emphasized that development was strongly buffered against both extrinsic (environmental) and intrinsic disturbance. Waddington developed a "canalized landscape" metaphor or model of development that is by now extremely familiar. He envisaged developmental pathways as a series of alternative valleys cut in a landscape. Early in development, decisions are made that direct the course of development (represented by a ball rolling down the landscape from high ground to valley floor). The steep sides of the valleys represent the extent of buffering, so that if the ball is deflected to one side or another, the properties of the pathway (slopes of the walls) bring the ball back on course. Some pathways are more canalized than others (deeper versus shallower valleys). Within any valley there is a range of possible phenotypic expression (where the ball comes out at the end of the valley), but the range of choices is tightly controlled. It is a concept that is almost self-evident, but if it is to work at all it must require great redundancy of information, perhaps beyond our current capacity for technical analysis.

There is considerable circumstantial evidence for canalization, in that the response of phenotypes to a range of environmental or genetic perturbations is often not linear. For example, Rendel (1967) has analyzed one such case, the canalization of bristle number in *Drosophila*. The wild-type condition for *D*.

melanogaster is that there are four scutellar bristles. Under various combinations of the recessive mutant *sc*, bristle number is modified. Rendel analyzes the situation by envisaging a pool of an hypothetical factor "make" that is produced by a range of major and minor genes under various controlling influences. The *sc* mutant affects levels of make, but the response of bristle number to hypothesized make level is nonlinear. A wide intermediate range of make values all produce four bristles. At low and high make values, only a small difference in make level will cause a change in bristle number. This is obviously heavily tied up with the phenomenon of dominance. Under strong selection it is possible to alter the response to make levels, and populations can be selected that become canalized for two bristles rather than four.

There are two principal objections to this sort of analysis. The first is that it requires postulating a purely conjectural factor "make" for which we have no direct evidence. Second, it tells us nothing about the dynamic properties of the canalizing system—its development (which is after all what the concept is all about). In Rendel's scheme, canalization works through tolerance of a particular range of disruption caused by the expression of the mutant *sc*. In a truly buffered situation, if expression of *sc* disrupts the system, other agencies must react to allow the morphogenesis of bristles under the new conditions. In a buffering situation, the developmental pathway would react by producing the normal amount of make. In the latter case, it is necessary to have a fuller model of the developmental control of bristle number than the idealized notion that there is a pool of something called "make."

A possible demonstration of partially unsuccessful canalization may be provided by the group of mutant limb conditions in mouse known as luxate. As Hinchliffe and Johnson (1980) review, the phenotypes have in common a reduction of the tibia and a tendency to polydactyly. They suggest that the problems are caused initially by the limb bud being excessively small, smaller than can be compensated for by increased rates of cell division. There may also be a deficiency in the preaxial part of the AER. The result is an insufficiency of mesenchyme for the prospective tibia region. Eventually this can lead to loss of the preaxial digits. However, the preaxial AER may be prolonged later into development rather than regressing normally, and when this happens there is a compensation in the amount of mesenchyme available. This is too late to rescue the tibia but can rescue the digital arcade. However, what may also happen is that too many digits are then formed from this extranormal mesenchyme (polydactyly).

The important feature of this last way of looking at canalization is it concerns morphogenetic rather than cytodifferentiative mechanisms and thus concerns the rates of doing things rather than the number of things being done. This makes them harder to analyze. Unfortunately, experimental manipulations of developing systems have not yet yielded many results here. For example, it has not yet been possible to reproduce through experimental manipulation of AER size (by grafting or excision) the range of effects that is seen in mutants where AER size is changed.

Waddington used his discussions of canalization mostly to exemplify the resistance of pathways to change. But he also highlighted their special potential for change in that various factors could push the course of development from on pathway to another. Selective or nonselective changes can change the shape of the canalized landscape so that, putting it very simplistically, if the new pathway is favored, those genetic changes that reshape the switchover point between the two pathways will be favored and the canalized landscape reshaped and recanalized. A feature of the topography of the landscape will be that some areas are more strongly canalized (deeper valleys) than others and so less labile and open for change. Once a particular pathway is strongly selected for, then associated genetic effects that increase the canalization of the pathway will also be selected for through selection of phenotypes that are minimally disrupted and optimally "efficient" in developmental terms. The result will be that the particular pathway is always chosen and then strictly adhered to. Thus we have to be careful to distinguish between two meanings of the term "canalization." In the first it is a state of development; in the second it is the evolutionary process that produces such a state.

As we have already noted, it is difficult to translate the metaphorical valleys and balls of the canalized landscape model into empirical developmental terms, but Waddington's model does have one very important property in that the changes forcing the course of development from one mode to another are all threshold phenomena (see, particularly, Rachootin and Thomson, 1981; Thomson, 1982). At the point of switching between pathways, whether we are considering the normal course of development where a presumptive tissue becomes restricted and divided into subregions with separate fates (regionalization of mesoderm), or whether we are considering an evolutionary change in pathway (such as induction of neurectoderm), the phenotypic effect of an extremely small initial shift could be very large.

This produces a potential paradox of great importance to evolutionary considerations. The phenomena and processes just discussed will tend to uncouple genetic and phenotypic variation. When discrete differences exist between the phenotypes of organisms in different populations or species and the switch between the two conditions appears to be triggered by a single gene, the difference could represent one of two very different situations. It could represent the directly expressed effect of a single allele substitution (perhaps eye color in *Drosophila* works this way), or it could represent the accumulation of very large numbers of genetic differences affecting the particular condition but previously canalized until a threshold was crossed by addition of one extra substitution. In this case, the last substitution does not "cause" the phenotypic difference. (This is different from a discussion of genetic backgrounds, which would also be relevant to any change, as it concerns the genetic control intrinsic to the particular pathway.)

Another paradoxical situation related to canalization of phenotypes is phenotypic variation. A population in which a particular phenotype is highly canalized will show low phenotypic variation, but this is likely to have been achieved by means of accumulation of a large number of genetic "adaptations" in the form of fixing of alleles favoring canalization. On the other hand, high phenotypic variation in a population in certain cases might be due to lesser canalization of the pathway, as fewer genetic modifications of the pathway have been accumulated. Once again, the quantity of expressed phenotypic variation may be an unreliable index of the amount of genetic change involved.

But we still do not have a good explanation of what canalization actually represents in terms of developmental mechanisms. It will be very different in morphogenetic pathways compared with biosynthetic pathways; we will concentrate on the former. As discussed previously, morphogenesis is a sequential process possibly involving three related phenomena: a cascade of contingent, interactive signals; the buildup of complexity in a geometric progression; and the operation of highly specific pattern-generating systems. All work through systems of signal and response and the control of properties of cell behavior. But while these cell properties, especially early on, are very generalized, each presumptive cell type must be significantly different in terms of the numbers and types of genes that are active at any time. It is these differences in genetic expression that allow each cell type to signal to its neighbors in the constantly changing epigenetic nexus of signal and response. How can these mechanisms be buffered or canalized?

Taking the example of the limb bud (see luxate mutants, above), we can see that the result of early morphogenesis under the patterning influence of the ZPA and AER is to apportion mesenchyme to various regions of the limb bud. A range of errors made at this early stage with respect to control of initial numbers of cells can be accomodated later by change in cell division rates or recruitment rates within the apportioned mesenchymal region, or, alternatively, changed rates of cell death. These will produce blastemata of presumptive limb bones at "corrected" size. Supposing that accommodation at the preblastema stage is not completely effective, a smaller range of correction is possible at even later stages through alteration or rates of cell division within the blastema, by rates of division of chondroblasts at the cartilage stage, or even by cartilage hypertrophy through swelling of the cartilage matrix. Finally, even after birth, control is possibly over the rate of growth of the bony elements and plastic remodeling is possible throughout life. At each stage the possibility for correction becomes less.

Some perturbations of the developmental pathway are of sufficient scale or of such a particular nature as to be unavailable for correction in later stages. For example, in luxate mutants the tibia cannot be rescued by later AER effects. And at later stages, if muscles attach to the wrong blastemata or the pattern of innervation is faulty, this cannot be corrected for at the local morphological level. In general, where the morphogenetic mechanism concerns the production of quantitative differences, the developmental system has a certain capacity to accommodate error if the canalized pathway is one in which size and rate phenomena are under the control of mechanisms that respond to (and thus in some way monitor or affect) whole-organism properties. Whole classes of qualitative phenomena will not be under such control. For example, it may well be that the matter of whether two separate blastemata or one fused blastema will form in a given region of the carpus is basically a qualitative phenomenon. The fusion of two blastemata may be due to passing a threshold in terms of the physical separation of the two. But once the two have fused, no future process involving the formation of a chondrification and the replacement of that cartilage with bone will be able to reverse the error. A qualitative shift has occurred. Some of the most crucial stages in pathways must therefore be the decision points that produce qualitative differences in the fates of cell lineages. In regulative development, as in vertebrates, these will include especially the points of "instructive" tissue interactions.

A canalized developmental pathway is therefore one that has the capacity for self-correction, for the accommodation of genetic/developmental perturbation. This can occur most readily in early stages of morphogenesis through compensating rate phenomena within the pathway.

Canalization can be selected for in populations in order (as Rendel puts it) to optimize "a balance between developmental processes" (1967). Canalization in this case is essentially a consequence of stabilizing selection, with very high selection against any genetic variant that would bring a pathway to a threshold level for disruption and concomitant selection for genetic constitutions that are conformable with maintaining the functioning of the whole. Strong directional selection will tend to disturb the canalization of developing systems, and any new phenotypes are likely to be unstable until recanalization of pathways has occurred. An example of this phenomenon has been given in a study by Clarke and McKenzie (1987) showing increased frequency of morphological asymmetry in blowflies exposed to pesticides, followed by decreased frequencies after continued (canalizing) evolution of populations under pesticide exposure. Canalization is particularly likely to be disturbed by inbreeding, which reduces the genetic diversity of the control mechanisms. This is possibly the cause of variability in Hanken's peripheral population of *Plethodon,* discussed above. Interestingly, Rendel, 5 years before Eldredge and Gould first published on "punctuated equilibria," observed that the phenomenon of canalization is likely to lead to patterns of evolution in which "periods of great evolutionary change will be short-lived and will be interspersed by long periods during which nothing much seems to happen" (1967).

Constraints

The buffered, self-correcting properties of developmental systems lead us to consider a closely related general property of developing systems, that of developmental "constraints" (review in Smith et al., 1985). Although canalization as a state allows things to work when some of the bits and pieces are out of synch and canalization as an evolutionary process allows for accommodation over time to such perturbations, nonetheless such capacities are finite and distinctly limited. Development in complex organisms involves such huge numbers of genes, such complex cascades of defining mechanisms, and such a diversity of cell and tissue types, all starting from a single cell in which the genome is not yet switched on, that development is always tightly "constrained." That is to say, not only is development a strongly normative process aimed at producing a given result; the number of ways and the types of ways in which development can be changed to produce other results (evolution) are also strictly limited. The term constraint applies to those limitations that have an intrinsic cause, although as we shall see, intrinsic and extrinsic processes necessarily combine in practice. The very self-controlling properties that allow regulation and canalization depend on a very large range of "givens." Developing systems have a series of built-in properties, patterns, and mechanisms that cannot readily be altered.

The restrictions of potential change that developmental constraints represent have multiple, complex causes. We have just discussed the first of these, namely the very interactive complexity of developmental cascades (which we have termed buffering) and its dynamic reinforcement, which is canalization. However, canalized pathways can be changed if the right "impetus" is delivered and a threshold is crossed that effectively reshapes the developmental landscape. The term "constraint" is more usually expressed in terms of two other properties of developing systems: the constancy that is inherent in the nature of "developmental rules" and the inertia that is expressed in the accumulation of history in the evolution of morphogenetic systems.

Developmental rules operate at any level in development. Allometric relationships would be a simple example of late-operating rules. The relative size of the antlers of all known species of deer reflects a single allometric relationship (Gould, 1973). The pattern of folding of the sulci of the cerebral hemispheres of mammals can be related, by two allometric equations, to adult brain volume and a relationship between brain volume and surface area. Whatever causes these allometric relationships forms a constraint. More difficult to break or bend would be deeper pattern formation mechanisms, such as those controlling the patterning of the tetrapod limb (see below), or even deeper aspects of pattern such as symmetry. The boundary conditions of developmental rules are also factors that are historically contingent—cell type, nature of extracellular matrices, and so on. These properties are also highly constraining in terms of possible evolutionary change. "What might be" is obviously a function of "what has been" and "what is."

One of the classic examples of study of developmental constraints has concerned the tetrapod limb. As Holder (1983a) and others have pointed out, in the actual manifestation of tetrapod limb types over the 300 million years of its history, there has been not only an extraordinary consistency in what has been produced but also a consistency in what has not. If developmental consraints are real, then certain morphologies should be impossible to produce in either nature or the laboratory. Holder has reviewed the history of the tetrapod limb and recorded the so-called forbidden morphologies (Figure 12). There is, for example, no known case of a limb with digit 3 reduced and the others retained. These patterns of tetrapod limb evolution match very closely with what is known of the developmental control over limb structure and the various models of limb pattern formation and limb evolution, which are, as already noted, based on the same data anyway.

The simplest and most obvious element of constraint in this system is that

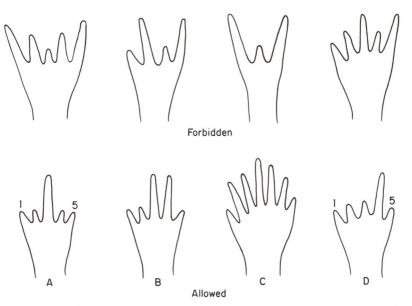

Forbidden

Allowed

Figure 12 Some "forbidden" and "allowed" tetrapod limb morphologies. 1–5 = digit number; A = Perissodactyl, e.g., modern horse); B = artiodactyl (e.g., modern deer); C = icthyosaur; D = flying reptile (pterosaur).

where there is a strong single axis of organization, the medial digit will be the hardest to lose. If digits are lost, they are more likely to be lost from the preaxial and postaxial margins of the limbs, as in the horse evolution sequence (first losing digits 1 and 5, then 2 and 4). It also follows that where the digits are formed in sequence, in some forms of extremely simple heterochronic perturbation of the system, the last digit to be formed will be the first to be lost (see Alberch and Gale, 1985).

However, there are many known exceptions to these simple formulations. And there is no easy answer to the adaptive argument that runs thus: if there were ever strong selection for reduction of the middle digit and retention of the lateral ones, that pattern would be produced. That it is not seen may only mean that the particular "forbidden morphology" is presently functionally inadaptive and has never been called for, rather than being developmentally excluded. Lacking sufficiently detailed developmental analyses to attack the problem directly, one approach has been to use computer models such as those of Goodwin and Trainor (Chapter 6) or Ede (e.g., 1976). These models have the advantage that given mechanisms for generating morphogenetic patterns can be modeled and then systematically modified and the results visualized. But the model will only give out results according to the information that is put in. If the premises of the model inherently forbid certain morphologies, then naturally the results of manipulation of the model will be to confirm that these are never realized.

Most obviously constrained aspects of morphology must be under controls that are both intrinsic (developmental) and extrinsic (functional-adaptive). For example, the shapes of individual bones in the skeleton must be materially constrained by the fact that they are made of bone or cartilage. The main reason why, for example, the human femur can never be 30 times as long as it is wide is functional-adaptive. A bone that shape would easily break and would be impossibly unwieldy as a lever in a jointed-limb system. In practice, selection constrains the shape of the femur in *Homo sapiens* to within a particular range, and selection has favored those gene arrays and mechanisms that will generate such a range. This is therefore not just a matter of fundamental developmental costraint; rather, it is an adaptive constraint developmentally expressed. On the other hand, it does represent an historically contingent constraint in that successive populations of *H. sapiens* inherit the same gene arrays. The controls could be modified through further selection, but then the question becomes, how, and to what extent? To what extent is femur shape limited by local controls (concerning the femur only) and to what extent is it bound up with the general pattern control mechanism of the whole limb and thus inescapably bound up with the relative size and shape of all the other limb components, skeletal and nonskeletal? This is the sort of question to which we will return later in considering the control of morphological characteristics.

Significance and Consequences of Developmental Constraints

Two aspects of developmental constraints are extremely important to evolutionary sciences. First, the general features of any developing system will be constrained in the sense that they can only be changed with great difficulty. Second, it will be a particular property of any constrained system that, to the extent that it is alterable, it will be more readily changed in certain directions than others. That is, developmental constraints are likely to be important among the factors that both limit and shape the range of potential phenotypic variation. Because the properties of constrained systems are carried over from taxon to taxon in evolutionary time (the more generalized systems being held in common by greater numbers of taxa), developmental constraints can potentially have long-term importance and contribute to the explanation of such puzzling matters as evolutionary trends.

Developmental constraints explain the "phyletic constraints" that have been part of the vocabulary of evolution for a long time (see, for example, Gould and Lewontin, 1979). Phyletic constraint is the property of a clade of which all members share the same developmental constraints and express them in a similar phenotypic fashion. For example, the fact that all branching structures are formed in vertebrate morphogenesis from epithelial organs is an example of a phyletic constraint. Other animals form them perfectly well from mesenchyme. The limited number of mechanisms by which epithelial organs actually form branching structures represent fundamental developmental constraints. Phyletic constraint is classically demonstrated in the case of the evolution of the tetrapod limb, just discussed.

At the very least, developmental constraints must stem in part from the complexity of developing systems and their control, and from the strongly conservative influence this complexity must have over possible change in pathways over evolutionary time. Whatever has been in history will tend to be conserved. This is the basis of phyletic constraint and is the basis on which we recognize homology. Throughout the whole vertebrate subphylum a particular way of making a branchial skeleton through neural crest cells and an epithelial interaction in the pharyngeal region is strongly conserved, presumably because it is too complex to change (constrained). The result is that it is possible to argue that the epibranchial element of the skeleton of the second visceral arch of fishes is the same thing as (homologous with) the stapes in mammals. And despite current controversy concerning the precise homology of the mammalian stapes with structures in lower vertebrates, no one any longer thinks that, for example, Gegenbaur was right that the pectoral limb skeleton of vertebrates is homologous with any part of the visceral arch system (Goodrich, 1930).

Developmental constraints also explain two additional phenoma: apparently vestigial embryonic structures such as the evanescent formation of branchial pouches and slits in mammals, and the unbreakable correlation of parts in morphology. For example, gill pouches are formed in mammalian embryos, although no gill structures are needed. The embryonic pouches are formed, not because of the operation of a recapitulatory principle, but because all vertebrates assemble their pharyngeal morphology in the same way. These mechanisms are fundamentally linked to other mechanisms acting at the same time. The formation of these paired pouches in turn is a controlling influence in the morphogenesis of the visceral skeleton, the aortic arch system, branchial musculature, and associated endocrine glands. Successive modifications of these structures, seen in historical sequences of phenotypes, have occurred through modification of the same basic set of morphogenetic processes. The occurrence of what appear to be "fishlike" branchial pouches in advanced vertebrates therefore represents nothing more than the latest version of the same set of mechanisms, modified in accordance with producing the latest manifestation of the original pattern of morphology. Thus the mammalian embryo produces no more "vestiges" of the original vertebrate morphogenetic pattern than it needs to in order to assemble the homologous branchial structures present in mammals. At the same time, in the history of mammals, new structures (endocrine organs) have taken over the pouches, giving a specific reason for their conservation.

The same phyletic developmental constraints explain the morphological correlation of parts across wide ranges of taxa within a major group. For example, there exists a remarkable constancy of correlation of parts in the ear ossicle region (again part of the visceral-branchial skeleton). It is possible to show that the mammalian ear ossicles are homologous with portions of the gill arch skeleton of fishes. Across the whole range of vertebrate types, from amphibians with only one ear ossicle (stapes) to mammals with three (malleus, incus, and stapes), homologous structures maintain the same pattern of morphological connections because of their fundamental developmental connectedness. In the jaw–ear ossicle complex, this connectedness has been experimentally tested by

Kay (1986), who used retinoic palmitate to disrupt morphogenesis of the first and second arch complex in mouse. Kay showed that the results were correlated by the arch unit to which the structure belonged, mandibular versus hyoid, etc. This is an important principle not only in attempted analyses of homology but also analyses of the patterns of change, for it means that changes in one part are bound to have consequences for other parts and, mutatis mutandis, the range of potential change of the whole complex will represent a major constraint on evolution, and possibility for change of any one part. This is a developmental constraint that reflects both what is developed and the morphogenetic mechanism concerned.

The major significance of the concept of constraints to evolutionary analyses is that the rules of developmental processes, especially as manifest in developmental constraints, directly and significantly affect the expression of phenotypic variation. We have already noted that the canalized properties of pathways may cause them only to change through threshold events, uncoupling genetic and phenotypic variation. In addition, developmental constraints may uncouple the two in a different manner. It may even be the case that the rules of pattern control in a given system may be such that a whole variety of different genetic perturbations will cause the same range of phenotypic effects. For example, a number of quite distinct genetic variations could cause identical phenotypic changes in the size and shape of the tetrapod lim bud (as we can observe in comparison of different mutants). Thus, however produced, various modifications of limb bud shape will have the same small range of phenotypic effects—for example, the various syndactylies and polydactylies seen in mouse and chick mutants (see Gruneberg, 1963 and Hinchliffe and Johnston, 1980, for review).

Lacking direct developmental analyses, we can use the typical evolutionists' device of a scenario to exemplify how such factors might affect evolution. In the case of the evolutionary transition from five to three to one toe in horses, a population-level explanation would be that selection strongly favored small phenotypic differences in the size of the lateral digits, which accumulated until digits 1 and 3 had been reduced beyond a functionally critical point (threshold). This then continued until digits 2 and 4 had also been suppressed. This assumes that selection can act on genetic differences whose effects slowly accumulate, passing certain thresholds, to be sure. However, a simpler developmental explanation is available. It is possible that strong selection in the early horse lineage for change in proportions of the three segments of the limb produced requirements for the apportioning of mesenchyme within the developing limb bud (longer blastemata of the zeugopodium and autopodium) that could not completely be achieved in the earliest stages. This could occur through sorting of variants in the shape of the early limb bud and in the mode of AER regression. Reapportionment of mesenchyme away from the digital to autopodial region then occurred such that the blastemata of the lateral digits failed to achieve critical size and were eliminated before the chondrification or ossification stage. Any allele substitution favoring reprogramming of the allometry of the limb segments would therefore directly contribute to the suppression of the lateral digits. If the resultant set of phenotypic modifications turned out to be functionally useful (adaptive), as evidently was the case, all this could be reinforced by whatever allele substitutions helped the general cause.

In this we may even assume that many correlated changes in limb structure and function associated with loss of lateral digits and unguligrade locomotion will follow directly mechanically from change in skeletal morphology (as in Muller's work on the chick limb, 1986). In this way a large range of interconnected changes in muscles, nerves, ligaments, and shaping of joint structures can be accounted for without the necessity of invoking separate selective pressures on each system. The phenomenon of genetic assimilation (see below) would be crucial in reinforcing and stabilizing such correlative effects.

The significance of such a scheme is threefold. First, it is developmental constraints (the rules and boundary conditions) of limb development that control the pattern the digit suppression and modification of the carpotarsus. Second, phenotypic variation is introduced into the population in a nonsymmetrical way. That is, a whole range of musculoskeletal variants cannot occur—not because genetic variations coding for them do not crop up in the genotype but because the pattern control system has its own rules controlling phenotypic variation. Third, the genetic changes involved all concern the early stages of limb bud morphogenesis, rather than late phases of gene expression.

A further property of this model scheme is that the transition from five to three to one toe can potentially proceed at a fast rate because it is not dependent on the temporal summation of thousands of very tiny phenotypic changes in digit size, but on rather simple changes in fundamental rules. We will expand on this theme in Chapter 9.

If we look at developmental constraints from the opposite perspective, we see that not only do constraints literally constrain (i.e., they reduce the scope of phenotypic variation), developmental constraints may also dictate that certain morphological features are more labile than others. For example, in general it seems to be the case that in the tetrapod limb the proximal and distal regions are more tightly constrained than the middle (carpal and tarsal) regions. Thus Hanken's study of a variable population of *Plethodon cinereus* (1983b, see Chapter 6), revealed that the carpus and tarsus were more variable than other regions. Constancy of the humerus–femur or ulna–radius/tibia–fibula complexes may be inevitable because they involve fewer component parts, or are fixed early in development. But this does not explain away constancy of phalangeal pattern in this population. Then again, within the more labile carpal and tarsal regions, Hanken found that certain fusion combinations of elements were more likely than others, and a whole range of fusions that were topographically possible were apparently forbidden.

If the more common fusion patterns in Hanken's population are truly a product of developmental constraints rather than the accidents of genetic reshuffling in an isolated and presumably strongly homozygous population, they should occur with similar frequency in other isolated populations of the same or related species. Suitable intraspecific comparisons are not yet available. Cross-specific comparisons within the family Plethodontidae and between these and the family Bolitoglossidae (D. B. Wake, 1966; Alberch, 1983), however, partially support the conclusion that the pattern of variation is phyletically conserved.

The consequences of the nature of developmental pathways, their buffering, canalization, and constraints add up to a very simple fact about development. When a series of complex controls is imposed upon an initially homogeneous system, the result can only be a discontinuous patterning in both space and time. One might change Waddington's landscape metaphor to one of a pinball machine. The rolling ball of development may be pushed around by various forces acting within the system, but the available end points are fixed and nodal rather than infinitely extensible. Evolution will change the shape, size, spacing—even the number—of the cups into which the ball may fall, but they will remain distinct.

Atavisms

Related to the subject of developmental and phyletic constraints is the phenomenon of atavisms. Classic cases would be the occasional appearance in modern horses of a three-toes foot or the well-known bithorax complex in the otherwise dipteran *Drosophila*. Hall (1984) has made a thorough survey of all the known examples. By definition, all these mutant conditions express a phenotype that is assumed to have existed in the ancestry of the taxon in question but which has not been expressed normally for perhaps millions of years. They represent a sort of unraveling of development and particularly a sort of failure of the developmental cascade at particular points. As such they are powerful evidence for the fact of evolution and are sometimes thought of as evidence for recapitulation. They are never thought of as representing the addition of new genetic information, hence the name atavism—throwback. Instead they are thought of as representing the failure of a layer of genetic control.

If ontogeny were simply a consequence of phylogeny, as the old recapitulation theory proposed, then the range of possible atavisms in any taxon would be predictable from its ancestry. In fact, however, in principle one can only predict atavisms completely from a knowledge of developmental pathways and mechanisms. Thus, for example, homeotic mutants are known in *Drosophila* that produce atavistic conditions not known or possible in any ancestor. (One can duck this issue, of course, by saying that these are merely monstrosities.) A simpler case would be where changes to developmental pathways have obliterated information that was once present.

Perhaps the crucial point about atavisms concerns their relationship to "progressive" evolution (as opposed to minor variations on a theme of the sort discussed in the following chapter). There seems to be no evidence, and no reason in theory, that atavisms could represent a major opportunity or set of mechanisms for evolutionary change. In the history of some groups, there may have been one step backward before two forward. But atavisms are most interesting for dissecting out developmental pathways.

CONTROL OF PHENOTYPIC CHARACTERS

A central question in evolution is, how do major new morphological configurations arise? What is the relationship of the sort of normal-scale phenotypic
variation and genetic variation that one can observe among individuals in populations or between species to the nature and scale of differences that one ob-
serves between major groups? It is obvious that major groups differ in terms
of major features of their morphogenetic systems. What have been the pattern
and the process by which such differences have been arrived at in evolution?

As we have discussed, change in the nature of gene expression, whether in
a structural gene array or in genes controlling the expression of other genes,
can affect developing systems at any point once the zygotic genome is activated. Prior to that, maternal effects can modify the pattern of development.
Because the greatest number of genes are activated at relatively early stages of
development, the potential for perturbation of development is greatest there.
But this may be offset if, as is widely thought, the genetic coding of development includes a large amount of redundancy, and by the self-regulative powers of canalized systems.

When the genotype of an individual or population is altered, the effect in
terms of the phenotype will depend not simply on the nature of the causal
change (what the change in gene expression actually is) and its epigenetic consequences, but also on where in the developmental sequence(s) it is expressed.
It is a fundamental property of development that the major morphogenetic patterns are set in place earlier in development. They must involve interactions
with a more diverse array of other developmental pathways than more minor
morphological characteristics. In general, any perturbation to such early-stage
morphogenetic processes is likely either to produce no effect (canalization) or
a discontinuous effect (threshold crossed). Later-stage changes are likely to
produce smaller (graded) effects with a greater chance of being expressed. This
will apply in any developmental system, whether we are discussing the embryo
as a whole or the developmental pattern of some constituent part (e.g., limb
development).

This whole situation is at the basis of the classic developmental laws of
Meckel-Serres and von Baer: the hierarchical levels of morphogenesis corre-
spond to the setting in place of morphological features that characterize different major taxonomic levels. In general, the earlier in development, the higher
the level of taxonomic character that is caused and vice versa, although we
must always be careful not to fall into the trap of organizing characters by their
presumed phylogenetic history, but rather more directly by the points of their
cause and expression in development [Medawar (1954) points out that in vertebrates, it is only from the neurula stage onward that developmental stages
look similar. This is presumably because of specializations of the egg, principally concerning the yolk, that affect the blastula stage.]

Once again, therefore, we have to use hierarchical terms if we are to discuss
the control of phenotypic characters. We already have the outlines of five major

hierarchical levels of development: early pattern formation (two stages: maternal genome control and zygotic genome control), late pattern formation (early and late morphogenesis), and cytodifferentiation. We do no yet have enough information to make a more sophisticated hierarchy, but this will serve to establish the main points of the argument. In principle we should be able to reconstruct for any species or any higher group a sequence of levels of morphological characteristics that define all the higher groups to which the taxon belongs, and to match these up with particular points in the hierarchy of morphogenesis.

This kind of hierarchy sets the general context of development within which, over time, the developmental pathways of a given lineage will accumulate and fix endless variations, duplicated and edited, twisting and turning as it were (even doubling back perhaps) through a maze of gene expression.

Perhaps the most fine-grained level of phenotypic character that one could imagine to have evolutionary significance would be something caused (and thus varying) at the cytodifferentiation level. In this category would fall anything that constituted a specific mate recognition signal (see Paterson, 1978). A minor late change to a biosynthetic pathway could produce a significant change in a pheromone substance and would not seriously otherwise disrupt the functioning of the organism. If it could spread in the population (see discussion in Chapter 9), it would lead to "instant" speciation. One can find many examples in groups of animals and plants of differences among species that depend on very small differences in color pattern, behavior (e.g., mating songs), timing of reproduction, or subtle morphological features.

Phenotypic features set in place at late morphogenetic stages basically involve modulations of pattern control mechanisms, because the basic elements of organogenesis have already been set in place during early pattern formation stages. An example of late morphogenetic control of a character would be the pattern-generating system controlling pelage, as discussed by Murray (1981). Changes caused at this stage would seem to be the basis of the differences among most closely related species of birds and mammals. The different morphological characters of the bill and feeding mechanics forming the basis of diversification of Darwin's Galapagos finches or the Hawaiian honeycreepers are probably also characters controlled at this stage of development. In fact, a large array of phenotypic differences involving size and proportion due to the operation of allometric and heterochronic mechanisms are caused at the late-morphogenesis stage.

More gross morphological characteristics are controlled at earlier morphogenetic stages. Obvious examples would be the control of tetrapod limb patterns at higher taxonomic levels. The characteristics of the modern horse limb or the intratarsal reptilian ankle joints are set in place by factors acting in this stage, as are characteristics like the mammalian versus reptilian jaw joints. In his study of cichlid feeding mechanics, Liem (1973) found that a major group of African cichlids was characterized by a novel arrangement in pharyngeal jaw musculature. This was probably generated in early morphogenesis.

Even more gross character states are caused at the early pattern formation stage. Examples would be the condition in hexanchoid sharks, where one or two extra gill pouches with more or less complete associated structures are formed as a result of modification of the very early patterning processes involving induction in the foregut (see Chapter 6).

We have already seen that we can make exactly the same sorts of simple distinctions if we look at hierarchical levels within a more restricted developmental system such as that of the vertebrate limb (dissected apart by analysis of mutants, Chapter 6).

We can now rephrase the question asked at the beginning of this section. If different scales of morphological structure are controlled at different levels in the developmental process, can changes accumulating at one level combine to produce changes at another level? Specifically, while events occurring at an early stage will have a potential upward-causing effect at all later levels, to what extent if at all can small-scale changes occurring at late stages of development accumulate to influence the causation of large-scale phenotypic characteristics that are basically controlled at early levels? To what extent can development be reprogrammed in the course of evolution so that developmental pathways change and significantly new phenotypes are produced? What do populational sorting (selective or nonselective) processes act upon and how does this affect developmental patterns?

Let us again take the simple example of loss of an element from the tetrapod limb. It could occur as a result of processes acting at the preblastema, blastema, chondrification, or ossification stages. There are plenty of examples to show that each of these stages can be the point of origin of element loss in one taxon or another. The ulnare of the chicken wing, as we have noted, is formed as cartilage but fails to ossify. Now suppose that in a related species the ulnare fails to form much earlier, perhaps at the blastema stage. Is there any logic to suggest that this second species has gone through a series of ancestral stages in which the ulnare was lost first at the ossification stage, then the chondrification stage, and finally at the blastema stage? I think not. If selection acted upon on gene arrays controlling the ossification stage, those arrays controlling earlier stages would be untouched. On the other hand, once an element were lost at the chondrification stage, there would presumably be some selective advantage to the gradual fixation of any genetic variants that "clean up" the housekeeping of morphogenesis by modifying the earlier stages as well. This would be difficult, however, if—as seems inevitably to be the case at least in the earlier stages of vertebrate morphogenesis—the preceding stages do not exist in isolation but are necessary to the interactions controlling the morphogenesis of other characters in that particular pattern (we will continue this discussion of "downward causation" in Chapters 8 and 9).

Where we are concerned with additions to the complexity of phenotypes, the situation is more clear. Suppose a new element in the limb skeleton appears, or a new arrangement of the original elements. Such modification of the phenotype could be caused by changes at various developmental levels. The only limitation is that, the more major the change (the greater the morphological scope and the higher the corresponding taxonomic level), the earlier it must be caused. New branching and segmentation of the skeletal blastemata are impossible at the ossification stage and unlikely at the chondrification stage. One could not form a new ossification in the limb without a new cartilaginous blastema, and thus not without a modified set of mesenchymatous blastemata and thus a basic change in the pattern control mechanism.

If this is true it means that in modifying developmental pathways the scale of evolutionary change is limited by the level at which a particular phenotypic variation is caused. This both limits and extends our view of evolutionary mechanisms. It limits us because it means that we cannot assume that numbers of small-scale late-development-stage changes can ever accumulate to produce major reorganizations of the morphogenetic pattern. It delivers the coup de grace to "terminal additions." However, it extends our view because it shows us a means by which major alterations to phenotypes could be accomplished, arguably over a shorter time frame and with less disruption of organismal integrity than would be involved in wholesale serial accumulation of small additions (Chapter 9), and it offers the possibility for asymmetric introduction of variation, the asymmetry being due to the various properties (constraints) of the morphogenetic system rather than to the underlying genetic variation.

We have already seen this in the scenario of evolutionary change in the limbs of horses. To assemble the suite of changes between *Hyracotherium* and modern *Equus* by the chance accumulation of mutations that individually reduce the digits (and, moreover, reshape the whole limb in a manner that is both functionally useful and compatible with the morphogenetic program) would presumably be an extremely lengthy business, even under very strong directional selection. However, in the scheme outlined on p. 88, all that is needed (and it is still against considerable odds) is developmental constraint with respect to limb proportion, such that change in proportion is linked to change in limb morphology (essentially an oligosyndactyly). Then if the appropriate genetic changes affecting relatively early-stage morphogenetic patterning of the limb with respect to size and proportions actually appear in the populations, *and* if the concomitant changes in limb morphology are either not selected against or positively selected for, then both the rate and extent of change in limb morphology will be limited only by the extent to which size and limb proportions can be changed without functional disruption (see earlier discussion). This whole discussion will be continued and extended in the next two chapters.

Genetic Assimilation

Essentially as a footnote to all this, we need to discuss genetic assimilation (another concept from C. H. Waddington). An important property of epigenetic systems is their continuity, from first cleavage to the death of the individual. In this there is a continuing interaction between the individual and its environment, following something like the pattern shown in Figure 13. Even the mature phenotype has a certain potential range for remodeling of quantitative expression of particular features. Under given environmental conditions, the

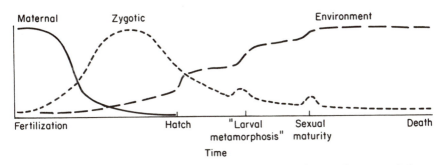

Figure 13 Diagrammatic representation of various contributions to the cause of change in a phenotype during life history.

phenotype may change. A simple example of this is the development of calluses in the skin of human hands and feet in response to mechanical abrasion. This response would not occur unless it had a particular genetic and developmental basis, and this is true of all such "acquired characters" as long as they are normal biological characters. It is logically possible, therefore, for such characters to be expressed in the fetal phenotype under a range of different controls, and humans are born with thickened skin on the soles of the feet. What presumably started out as an "acquired character" has, under strong selection, become an inherent character. This is genetic assimilation (Waddington, 1961).

Waddington and Thoday, among others, have studied this phenomenon experimentally in a variety of taxa. Perhaps the best-known case study is that of the condition cross-veinlessness in *Drosophila*. This condition is known from a mutation *cv* and can also be phenocopied by use of various shocks at a critical stage of development. Waddington heat-shocked lines of *Drosophila* and then selectively bred from those that responded by producing the *cv* condition. The frequency of *cv* in the population increased over time and after a while, when heat shocking was stopped, the population continued to demonstrate a high proportion of *cv* individuals. What had been summoned forth by an environmental cue had become inherent. In the reciprocal experiment, selection against *cv* reduced the frequency in the population and this again was stable after cessation of the experimental insult. Such genetic assimilation of an "acquired character" is one of the minor paradoxes of development in that it is both counterintuitive ("acquired characters" smacks of Lamarkism) and yet scientifically reasonable. The key is that the only characters that can be assimilated are "natural" ones—that is, ones that have a genetic basis. Cutting off the tails of mice (Schmalhausen, 1949) produces an acquired effect that cannot be assimilated because it has no basis in the normal genetic or developmental program of the population. A fine population-genetic explanation of genetic assimilation has been given by Milkman in terms of selection of alternative alleles that cause expression of a given character under different thresholds of environmental cue. A sort of genetic assimilation at the enzyme function level is perhaps shown by the case of "contingently dynamic genes" (Campbell, 1982).

Genetic assimilation is yet another example of a threshold phenomenon. It works by a shift in the balance of genetic/developmental and environmental factors necessary to trigger a particular phenotypic condition. It is interesting to evolutionists (Rachootin and Thomson, 1981; Ho and Saunders, 1979) because it shows the capacity of selection to reprogram development in order to alter phenotype expression within a given proscribed range. This range is of course limited by the genetic/developmental basis of the characteristics concerned. Genetic assimilation presumably cannot reprogram development in such a way as to produce characters more extreme than those that can normally be called forth by an environmental signal and—as Rachootin (personal communication) has put it—still means waiting around for alleles or combinations that "genocopy" the new phenotype.

8
Patterns of Evolution

We now need to look more closely at the evolutionary patterns that we wish to explain. Again, we can use the study of patterns to discover and define important problems to be solved. The methods available for the study of patterns depend on the focal level of the mechanism with which one is concerned. If our interest is in individual variation, naturally we need extensive sampling and laboratory experimentation, looking at individuals within populations. In order to study the biology of populations we must work at the population level, and at such phenomena as gene and character frequencies in the field and laboratory. It is at this level that workers have most readily been able to measure selection coefficients and other quantitative elements of evolutionary science. If our interest is in the processes of speciation we must look as closely as possible at examples of the process in action—at sibling species, at a whole range of hybridization patterns in the wild and laboratory, at the distribution of closely related species in space and time. (It will be noted that between population-level work and species-level work there is an unfortunate gap; one can look at prespeciation and postspeciation situations but it is rare indeed to be confident that one is looking at speciation *in flagrante delicto*.) We can also look beyond the species level at phenomena of species distribution within higher taxa. Here we progress to using the methods of systematic biology and comparative morphology and paleontology.

Each of these different approaches gives a different view of the evolutionary process and, therefore, in terms of the questions discussed in this book, gives us a different view of the role of mechanisms acting at the developmental level within evolutionary mechanisms. Obviously the most immediate element of causality in the origin of adaptive structures resulting from developmental properties is in the mattter of individual phenotypic variation, but some of the most interesting questions concern the consequences of developmental properties for higher focal levels–their upwardly causing properties. To examine these we need to look at the species level and beyond (what has often been termed macroevolution). We need to examine the pattern of diversification among species over the longer time course that produces those entities that we term higher taxonomic categories.

The problem can be phrased in terms of a question: are there patterns in the evolutionary diversification of populations and species that are the result in any part (by upward causation) of processes of developmental biology? Can we find developmental explanations for phenomena that are not fully explainable in terms of processes acting at other focal levels? In this chapter I will focus on two main problems. The first is the problem of coordinated evolutionary change in multiple characters. It is exemplified most acutely and has been studied most assiduously in the problem of the origin of major groups. When a major group of organisms arises and first appears in the record, it seems to come fully equipped with a suite of new characters not seen in related, putatively ancestral groups. These radical changes in morphology and function appear to arise very quickly, especially in comparison with the normal pace of evolutionary change within a given lineage. If real, how do such changes occur? One must find out as much as possible about the actual pattern of change, and then try to fit a mechanistic explanation to it. The second major problem is to discover some general properties in the patterns of evolutionary diversification within groups. Are there common patterns in the diversification of all groups, patterns that will reveal common properties of the causal mechanisms involved in diversification or, as it is often called, "adaptive radiation?" These twin problems of the origins of groups and diversification within groups can, of course, be studied at all levels, from the deme to higher taxa. A major task is to see whether there are differences in pattern and in mechanism at the different focal levels.

MORPHOLOGICAL INTEGRATION AND THE ORIGIN OF GROUPS

It is a well-known feature of all organisms, even the most apparently simple, that their structure forms a highly integrated whole. There are few characters or character states that can be considered completely "free" and unencumbered by functional-mechanical interactions with other characters. Study of such morphological integration at the organ and organism level has long been the special province of European theoretical morphologists, and the work of the schools of Seilacher, Dullemeier, and Reidl is especially prominent at the present time.

Morphological integration is a phenomenon that is expressed solely at the phenotypic level. It obviously has important parallels with the concepts and phenomena of developmental integration discussed in the previous chapter, but is fundamentally different. Morphological integration is in part caused by developmental integration, in the sense that morphology is materially created by development. But the rules of morphological integration are principally the rules of function and of materials.

Any evolutionary change in a morphologically integrated system must involve a complex feedback between integration of the phenotype in functional terms and developmental integration. To some extent all characteristics of a whole organism are interrelated and integrated and are a function of both all the particular properties of the organism and all the whole-organism properties.

The unraveling of the nature of the feedback interactions between the two constitutes a major challenge in evolutionary biology.

The skull roof of vertebrates makes an interesting example of an integrated system. The arrangement of the dermal bones in distinct patterns that are characteristic for taxa is evidently highly constrained functionally. Where we find a set of dermal bones firmly attached to a solid underlying neurocranial structure, the dermal bones often form a mosaic of simple polygonal elements, especially in lower vertebrates (Figure 14). Where the dermal bones transmit highly oriented forces due to the action of jaw musculature and the axial musculature inserting on the occiput, the arrangement of the bones is highly nonrandom. The key element is the orientation of the sutures. As shown in Figure 15, the sutures are arranged perpendicularly to the principal lines of force acting across them. This in turn largely controls the shape of the bones in a functional sense. The size of individual bones is probably under a related functional constraint.

How is evolutionary change effected in such a system of integrated dermal bone shapes? Because of the hierarchical nature of the developmental control of phenotypic characters (Chapter 7), we can imagine a number of different levels at which a variant condition could become expressed and fixed. For example, at later stages of development, differences in the growth rates of bone rudiments could cause minor changes in the arrangements of the bones and their sutures. This can be tested (within certain limits) by experimental extir-

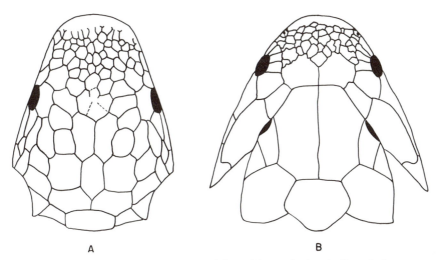

A   B

Figure 14 Comparison of the patterns of dermal bones in the skull roof of two types of Devonian bony fish: (A) lungfish; (B) holoptychoid rhipidistian. The skull roof of the lungfish is a mosaic while that of the osteolepid is a highly ordered pattern with unique bone shapes in different rent regions of the head. However, at the tip of the snout, where the dermal bones are fused to the underlying nasal capsule, the pattern in the osteolepid is also a mosaic.

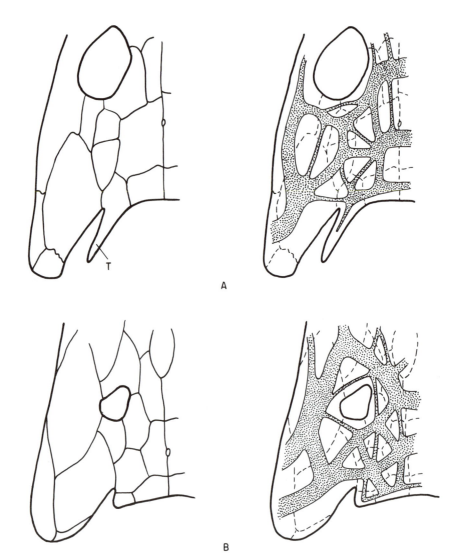

A

B

Figure 15 Comparison of the arrangement of the dermal bones in the posterior corner of the skull of two Carboniferous amphibians: (A) *Palaeoherpeton*; (B) *Eryops*. On the left the dermal bond sutures are shown and on the right the pattern of forces acting on the horizontal plane across the sutures is diagramatically represented. In *Palaeoherpeton* the cheek is not fully fused into the skull roof (hatched lines) and the pattern of forces acting in the bony elements is different from that in *Eryops*. In *Palaeoherpeton* there is also a prominent posterior horn on the tabular bone (T) presumably for muscle insertion.

pation of rudiments at late developmental stages and observation of the way in which other elements fill in. Phenotypic variation could be correlated with (driven by) small-scale changes in the mechanical arrangements of the underlying cranial muscle mechanics. At deeper levels of control, rearrangement of bone patterns might be caused by a change in overall skull shape driven in turn by change in shape or relative size of the brain. As du Brul and Laskin (1961) show, subtle changes in an unsuspected region like the basicranium may affect the patterning of the cranial vault. Such factors may be of particular significance in the rearrangement of skull morphology in mammalian evolution.

If the shape and relative size of each dermal bone in the skull roof were under separate developmental and genetic control, then to change the whole system would be very difficult and slow. However, if the growth program for size and shape of the brain and neurocranium is part of the boundary and initial conditions for the pattern control mechanism of the dermal skull bone arrangement, then complex and highly integrated changes in the skull roof pattern will come much more readily.

A second well-known feature of organismal biology is that the major groups of organisms (given by us the names of somewhat artificial higher taxonomic categories) differ from each other by suites of characters rather than single characters. These suites of characters are themselves usually suites of integrated characters. This integration takes two forms. First, it is functional morphological integration. Second, it is integration in the sense that, as we trace out the course of evolutionary change, particular assemblages of characters will change together over major periods of time, rather than separately. This gives an appearance of directedness to evolutionary change in morphology. Thus in horse evolution we find not only the structure of the limbs but also the head and dentition changing, apparently in concert. Where there are fewer digits there is also greater hyposodonty.

A third and related phenomenon has already been mentioned. It is that when any new group of organisms appears in the fossil record, it seems to possess a broad suite of defining characters for that group and that taxonomic level. The defining characteristics of a group do not seem to be acquired slowly and progressively during the history of the group, but rather several appear in combination. This is, of course, in part an artifact. We do not recognize the new group as a group until it has the full suite of characters that will characterize later members of the groups. We do not include the ancestor within the derived group anyway. Even though this phenomenon is in part an artifact, it has a germ of worrying reality about it; suites of characters seem to change together rather than single characters changing serially. It is probably this that accounts for the apparent sudden appearance and definition of new groups. Therefore, the linking of characters in change needs to be studied as well as the logic of how one defines groups. These three well-known phenomena help us to define a simple problem in evolution: what is the mode and mechanism of the origin of major groups of organisms?

One of the criticisms that is made of modern evolutionary study, with its concentration on the evidence from laboratory populational genetic studies, is that it is somewhat remote from the real world. In the laboratory it is difficult to study change in more than single characters defined by fixation of different alleles at a given locus. Similar problems apply to analysis of artificial breeding of domestic organisms. In the real world, especially at transspecific levels, differences among taxa tend to involve multiple characters. We are usually led to think that all changes involving multiple characters must have been caused serially, changes in separate characters being added one after another. But in this case, how does one get patterns of integrated change in multiple characters over geological time? If each change has to accumulate in a serial fashion, how fast can change occur? Is it possible that several characters could change at one? How do you get an integrated change in multiple major characters when all changes occur at the fine scale of the species level? Let us examine the pattern.

We will look at two principle examples: the origin of tetrapod vertebrates from fishes in the Late Devonian and the origin of mammals from reptiles in the Triassic. Both have been broadly studied for many years and in both cases we have an extensive (but who knows how complete?) fossil record, together with some good living analogues of supposed ancestral groups. Both schemes are no doubt heavily theory-laden, but they provide a useful example to start with. In both cases the time between any possible ancestor of the old group and the first member of the new one is relatively short, 10 million years or less. During that time, while closely related groups are only slowly diversifying in terms of morphology and taxonomy, the whole body plan of the divergent line becomes established. Then, once the new plan is in place, a new set of radiations is possible. The key question is: in what pattern does morphology actually change?

If we look at the two "ancestral" groups, Devonian lobe-finned fishes and Triassic mammal-like reptiles, we see several radiations of closely related sister lineages, each of which shows changes in one or more of the characters that together set apart the new derived group. But most only show a small number of such changes. For example, within the lobe-finned fishes, lungfishes show reduction of the gills in connection with increased lung function and they have a sort of internal naris. But they lack the tetrapod type of paired-fin and skull structure. Several groups of mammal-like reptiles show a type of primitive differentiation of the tooth row into specialized tooth types. This is close to what Schaeffer (1965) terms "experimentation." Probably all the related groups are experiencing the same general environmental conditions and can be construed as "attempting to" respond in the same functionally appropriate ways. As this phenomenon of parallel experimentation is particularly evident in the case of the reptile–mammal transition, let us examine this pattern in detail first.

The mammal-like reptiles arose in the Late Triassic, although the first fossils currently known all seem to be of Early Jurassic age (there are debates as to where to draw the boundary). Mammals are typically defined by having a jaw joint formed between the dentary and squamosal bones rather than the quadrate and articular. As we have noted, the latter two bones then come to form part of the middle-ear ossicle system. In fact the suite of characters defining mammals is broader than this. In skeletal features alone it includes jaw structure, jaw musculature, dentition (structure of teeth, replacement patterns, and tooth row differentiation), brain case and cranial vault, ear region, postcranial appendicular and axial skeleton. There is a corresponding wholesale overhaul of soft structures and physiology and behavior.

The mammalian condition is arrived at through a series of stages in antecedent mammal-like reptile groups, each stage being accompanied by a greater or lesser radiation at lower (included) taxonomic levels. Kemp (1982) has made a careful analysis of the sequence of acquisition of new characters, genus by genus within a phylogeny of mammal-like reptiles (Figure 16 and Table 1). This scheme is based on a particular phylogenetic hypothesis that may eventually come to be modified. However, the general pattern of change that Kemp has analyzed must in principal be correct.

In Kemp's analysis we can see that in the sequences of the taxa concerned there are certain points at which changes have occurred in whole suites of characters together. These include particularly steps G, J. L, and M on the cladogram (Figure 19). At each stage of the other stages there are changes in smaller numbers of characters. Of the characters analyzed by Kemp, the structure of the dentary and joint complex, the dentition, auditory system, and po-

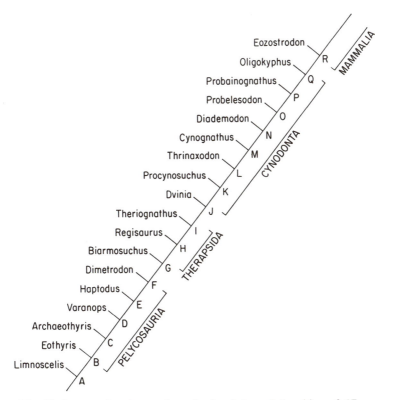

Figure 16 Cladogram showing an hypothesis of the relationships of 17 genera of mammal-like reptiles in the transition from reptile to mammal. Redrawn from Kemp (1982).

TABLE 1. Occurrence of Changes in Major Organ Systems During the Reptile–Mammal Transition[a]

	Dentary-squamosal joint	Jaw musculature	Tooth differentiation	Postcranial skeleton	Hearing-stapes	Brain size	Diaphragm	Nasal turbinals
A								
B		X	X	X				
C	X							
D	X	X		X				
E	X	X		X				
F			X					X
G	X	X	X	X	X			
H	X	X		X				
I								
J	X	X	X			X	X	
K				X			X	
L	X	X	X	X	X			
M	X	X	X	X	X			
N		X						
O		X		X				
P	X		X		X			
Q			X	X				

[a] Taxa A–Q as in Figure 6.

stcranial skeletal features are linked together at G, L, and M, but are unlinked at other stages. The most consistent linkage is between characters of the jaw mechanics, dentition, and postcranial skeleton: that is to say, not all jaw changes affect the skeletal apparatus associated with hearing. The linkage of jaw mechanics, dentition, and postcranial skeleton occurs at D, E, G, L, and M. The process does not end with the acquisition of those characters that we use for the formal definition of mammalness; there are three or four major different subgroups of mammals, and the accumulation of new characters continues until the placental or eutherian mammal condition is acquired. At this point all further radiation is currently seen as variation within the mammalian theme, rather than achievement of further new grades of "improvement."

A fascinating feature of mammalian history is that although mammalness is achieved by the end of the Triassic–Early Jurassic, and some seven or eight lines (or in some analyses more than a dozen—Norell, personal communication) must have been in place before the end of the Cretaceous, it was not until the Paleocene–Eocene that mammals started really to diversity at lower taxonomic levels. The reasons for this, as we will explore below, are probably extrinsic rather than intrinsic, but it raises the problem of how higher level innovations are, or are not, linked to lower level diversification.

In the case of the lobe-finned fish to tetrapod transition we have fewer data to work with and certainly lack anything like the sequence of generic transformations that Kemp has produced for the origin of mammals. However, we can see that there are the following characters that are labile within the whole lobe-finned group: skull proportions, jaw suspension, jaw mechanism, limb structure, gill and lungs, ear structure and function, nasal apparatus, postural control, rib pattern and vertebral structure, dermal bone structure, dermal bone pattern in skull, tail structure and function. In the different lobe-finned group we find various different combinations of changes in these characters all of which are change in essentially the same general direction. The are two rival schemes of phylogenetic arrangement of lobe-fins and tetrapods (Figure 17). I prefer the scheme according to which the first tetrapods must be thought to be descended from a primitive osteolepiform "rhipidistian." The crucial step in the origin of the osteolepiform–tetrapod line itself was the acquisition of an internally asymmetric paired fin, internal nares (choanae) possibly independently and in parallel with those in lungfishes, and a particular cranial mechanical setup that involves a specialized dermal skull roof pattern and jaw mechanics. Within the pretetrapod lineage, further changes in the jaw mechanics, coupled with loss of the gill apparatus (switch to lung breathing), and selection for a changed hearing mechanism, produces the elements of the tetrapod middle ear. As a result both of changes in cranial mechanics and loss of the intracranial joint system, and of elimination of gill respiration, the hyomandibular bone ceases to be a major jaw suspensory element and its role in sound transmission is increased. The first gill slit, no longer part of the respiratory system, becomes the middle-ear cavity. This is wholesale reorganization of the head. All of these changes occurred in one phase, without intermediate periods of diversification at low taxonomic levels.

Models of the Origin of Groups

Three basic models have been proposed to account for the acquisition of new character complexes in the origin of groups. The oldest is the "gradual" model, which states that the new set of character states is simply acquired slowly, gradually, and accumulates serially one at a time. In this model, the differences between groups are basically a matter of chance. The fact that the process of change seems to have proceeded faster than slow progressive accumulation of characters at the routine pace of gradual evolutionary change would seem to dictate must therefore be simply an artifact. Such a model requires very high taxonomic rates of evolution in order to give opportunity for enough small phylogenetically significant changes to be fixed within a given time frame.

The second model is the "key innovation model." According to this, change in one particular character is the key that triggers a suite of following changes, essentially by providing a sort of internal incentive for strong directional selective processes. An example of a key innovation would be the evolution of

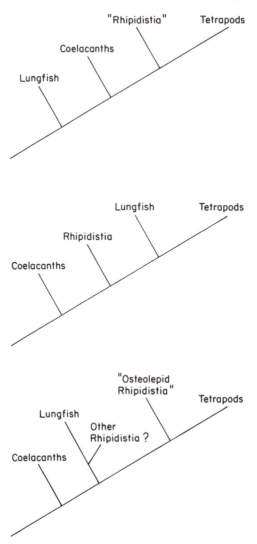

Figure 17 Cladograms showing three hypotheses of the relationship between the major groups of lobe-finned fishes and the tetrapods.

the feather. Once the feather had evolved, then the evolution of other features such as lightened bones and modified musculoskeletal system would have been facilitated in some way functionally (but not of course genetically or developmentally), and led to the origin of birds. A key innovation in the fish–tetrapod transition could have been the acquisition of lungs or the acquisition of a particular pattern of limb structure. There is a great deal of hindsight involved in picking out the key innovation. The key innovation model leaves a lot to be desired, mostly because it does not explain where the key innovation comes from and how it could become a fully functioning innovation without related changes already being in place. What use is a feather without wings, or indeed a wing without feathers? This is truly a "chicken and egg" paradox. It is necessary therefore to add to the key innovation model the eminently reasonable proposition of "preadaptation." "Preadaptation" (cf. exaptation, Gould and Vrba, 1982) is an awkward term expressing the sense that any innovation is really not new after all, but involves a change of function for a previously existing character which then, taking a new lease on life, becomes extremely important in the new mode. The feather again could be an example. There is a modest amount of evidence that something like feathers actually existed in the Triassic fossil reptile *Longisquama* (Sharov, 1970), in the form of heavily modified scales, with a possible function of providing a layer of external insulation for promoting retention of body heat. Their initial function was probably to assist in a primitive form of homeothermy. This fits what we know of the morphogenetic similarities between scales and feathers as modifications of integumentary structures (Oster and Alberch, 1982). The feather took on new functions when a group of reptiles started to experiment with jumping and gliding, and it went through a relatively quick transformation to the bird condition (continuing to function in endothermy). Similar arguments for change in function and thus preadaptation can be made for many other features (whether key innovations or not) in the origins of groups. It is an intriguing concept but still leaves open the question of how characters arise in suites rather than strings (see also discussion by Lauder, 1981, 1982).

The third model for the origin of complex morphological and functional character changes in the origin of groups is that of "correlated progression." I proposed this model in 1966 to account for the origin of the tetrapod middle ear from that of lobe-finned fishes. In the intervening period it has turned out that some of the data on which my account of middle-ear evolution was based are wrong, but the model itself remains useful and Kemp (1982, 1985) has used the same model in explanation of the origin of mammals (see also comments by Levinton, 1985). The concept of correlated progression depends simply on the observation that most characters do not—in fact, cannot—exist in isolation from each other, but are interrelated and integrated at various functional and developmental levels. If, therefore, any one character is subject to pressure to change, other characters cannot fail to be affected. When any character actually does change, that change will inevitably tend to be expressed in part as changes in other characters, although this will probably first be buffered or accommodated. There are obviously many characters in organisms that are not developmentally or functionally linked to others, but these will tend to be rather minor characters morphologically and (see next chapter) unconnected to other characters because they result from very late-stage gene expressions. Any part of a complex character state must be under control of multiple factors, and the causes of its change must therefore be various and complex, though not necessarily directly linked. A second part of this model is that environmental processes and selective regimes themselves affect multiple characters at the same time rather than single characters.

At the simplest possible level, one can imagine it would be possible to have a selective force acting to change some single component of an integrated system in isolation. Any change would then require compensating adjustments on the part of interrelated systems. In the case of the skull roof of vertebrates (Figure 15), we could suppose that directional selection to increase the posterior spur of the tabular bone in temnospondyl amphibians led concomitantly to change in the relative arrangement of a quartet of bones in the back of the skull table. The possible range of such change in a single component of a complex is bound to be limited by both constraints on changing the first character by itself and the constraining effects of the side effects on other characters. In this group of amphibians, the tabular horn happens also to form the anteromedial support for the tympanic membrane, which is part of the ear rather than the jaw mechanics. If the changes that jaw mechanics is "forcing on" the tabular are detrimental to middle-ear function, change may be slow or impossible. If the changes are neutral with respect to middle-ear function, change may occur, but still slowly. If, however, the changes are consonant with selective forces independently acting to improve middle-ear function (otherwise restrained by the nature of the tabular), then the combined effects will be to produce very rapid change in two disparate systems. Change in such complexes of characters will be faster, therefore, when there is selection on more than one component at the same time.

In the case of the origin of the tetrapod middle ear, I have suggested (1966) that changing mechanics of the jaw mechanisms (elongating the tooth row), caused associated changes in the braincase (fusion of the intracranial joint). These changes in the functioning of the head potentially made the hyomandibular bone unnecessary as a part of the suspensorium. In the parallel line of lungfishes, apparently this happened and the hyomandibular bone has become a small almost vestigial structure fused into the side of the posterior braincase. But in the tetrapod line there was simultaneous selective pressure to change the hearing function, for the reception of airborne sound, based on the preadaptive use of the spiracular gill cleft as a sound-receiving air bubble. Freed from its role in the suspensorium, the hyomandibular bone becomes the stapes and the spiracular gill cavity becomes the tympanic cavity. But all of this in turn would not be possible were it not for the fact that the prototetrapod lineage is also becoming more dependent on air breathing and the gill apparatus is reduced in size, leaving space for the increased volume of the middle-ear cavity. The developmental precursors of the opercular bone, which is lost as a gill cover, may be partially involved in the formation of the new tympanic membrane. In all this, as noted in Chapter 6, the fundamental connections of the stapes, its gill cleft, and the upper and lower jaws are maintained. Indeed this same fundamental connection can be traced to the mammalian condition where (it is argued) the stapes and two elements of the old jaw articulation become the three ear ossicles, still articulated with each other as they were in the first fishes, still connected to a tympanum, and still enclosed in the old spiracular gill cleft.

The correlated-progression model explains the origin of a suite of tetrapod characters from a lobe-finned ancestral stock or the suites of premammalian changes over the very short time frame. It requires that the ancestral forms be subject to strong selective pressures acting on a number of functional-morphological features at essentially the same time. These pressures are all connected with the adoption of a more terrestrial habit, and therefore there is a sort of "integration" of the selective regime at the whole-organism level. The response of the organism is itself integrated in the sense that developmentally and functionally integrated morphological systems are concerned. The gill system in fishes has functional connections to feeding mechanics, respiration, and hearing. The hyomandibular bone has mechanical connections to the jaw system, inner and gill apparatus. The visceral arch system is developmentally integrated in the morphogenesis of the whole vertebrate head. At all levels, change in one unit must be linked to changes in another. The whole suite of changes occurs when, so to speak, "all the arrows are pointing in the same direction." This happens rarely, but when it does, the effect will be rapid and coordinated change.

Not only will such change occur much faster in the coordinated-progression model, it may well be that this is the only model that will allow certain types of coordinated change. For example, if each unit of the whole were changed separately, change in one unit might proceed so far by itself, or have such consequences on other systems, as to prevent certain ranges of possible later change. The reduction of the hyomandibular bone of lungfishes might be an example. It was reduced separately before the rest of the gill apparatus; evolution of a tetrapodlike middle ear was then ruled out.

The argument against correlated progression is obviously that, if it is difficult to change any one major character, it must be many times more difficult to change several at once. Any monster produced by such discombobulation of genetic and developmental systems must be lethal, it is said. And how do you get such correlated changes at the population level? We will discuss these problems in Chapter 9.

A possible test of the gradual- versus correlated-progression models is as follows. The gradual model requires high taxonomic rates of evolution and low morphological rates. The correlated-progression model requires high morphological rates but does not require high taxonomic rates during the essential transition stages. Therefore, if the gradual model applies, we ought to find that major morphological change is preceded in the time sequence by major taxonomic diversification at the species level. In the correlated-progression model one would not expect to find any major taxonomic burst before the change (although it is not prohibited). Further, the coordinated progression model predicts (or at least allows) the rapid diversification of the descendant group both at the specific and transspecific level. Therefore, one should find a taxonomic burst not only at the species level but at higher taxonomic levels, following extremely quickly after the origin of a new higher level group. In the gradual model, such diversification should proceed only slowly. By these tests, in the origin of both tetrapods and mammals, the gradual model fails (see also Kemp, 1985; Levinton, 1985).

The Nature of Adaptive Radiation

Adaptive radiation is the term used to describe the diversification of a discrete taxonomic group in space and time and within a particular realm of functional-environmental contexts. The term begs the question of how much of a radiation is driven by the phenomenon of adaptation and stems from the time before the concept that all diversification was adaptive in nature became pejoratively re-labeled the "adaptationist program." In fact it is probably a fair approximation that most changes in diversity over a long period must be adaptive in the sense of being based on a morphological innovation that "worked" within a given environmental context. It certainly seems to be the case that most adaptive radiations involve functional characteristics.

Adaptive radiations are always finite in scope and by definition involve a phenomenon of increasing specialization and diversification at lower (i.e., in-cluded) taxonomic levels. This is always thought to involve, in turn, a pro-gressive narrowing of resource and niche breadth—a genuine specialization.

As just noted, we find that the course of evolution of major groups proceeds through pulses or stages of structural change, with certain stages being accom-panied by great success in terms of taxonomic diversification and distribution in space and time. Between these stages there are intervals where there may be considerable structural change but little diversification. We tend naturally to define groups by defining an envelope of diversification in a given mode. We also define groups in terms of between-group structural differences or gaps, which gives us a modest degree of control over our hindsight. The fossil record, at least in terms of major changes, does not always relate morphological and taxonomic rates directly; in fact they often seem to be opposed. This is quite different from rates within radiations where taxonomic and morphological rates usually match (because they are based on the same data). In fact, most div-ersifications concern morphological change at only a minor level (see next sec-tion).

What makes a group able to radiate or diversity? First of all, of course, there must be the intrinsic properties that make the group capable of succeeding, of surviving the general environment, and of dominating in the immediate envi-ronment of competition for resources. These intrinsic properties fundamentally make all members of the taxon relatively more efficient in the use of energy (see van Valen, 1976). Second, there must be the extrinsic opportunity. An example here would be the mammals. Mammals were modestly successful dur-ing the Mesozoic, but reptiles and other groups were as good or better, or simply got there first. Then conditions changed, perhaps principally the cli-mate, and it turned out that mammals had an advantage and could at last di-versify. The groups of early mammals probably were not so very different from the mid-Mesozoic forms, but the opportunity was different. A similar situation seems to have existed with the radiations of actinopterygian fishes. The te-leosts, which now dominate aquatic environments just as mammals do the ter-restrial ones, had their principal diversifications in the Cenozoic. However, the group first arose in the Early Jurassic or even Triassic, at a time when the chondrostean and holostean fishes that they eventually replaced were still undergoing their own radiations. What seems to have made the difference for teleosts was the opening up of new food resources in the aquatic realm, particularly small-sized planktonic and nektonic invertebrates (adults and larvae). Teleost feeding mechanics and locomotor systems turned out to be far better than those of their competitors in utilizing this resource. But they had to wait quite a while for the opportunity. The "opportunity" thus had both extrinsic as well as intrinsic components. The extinction of the older and formerly successful groups is obviously also partly due to some combination of intrinsic and extrinsic factors.

Once a group starts to diversify, that process tends to proceed at an exponential pace until a point of maximum taxonomic diversity is reached. In the groups that I have surveyed, this very high level of taxonomic diversity is not maintained for very long at all. Instead (Figure 18) the numbers of taxa inevitably starts an equally precipitous, essentially symmetrical, decline (Thomson, 1976, 1977). It is quite curious that we do not find many groups, if any at all, where high diversity continues for a significant period through the record once it has been achieved. The only exceptions that I know seem to be in the Mollusca. At least in vertebrates, where I have studied it in detail, the shapes of

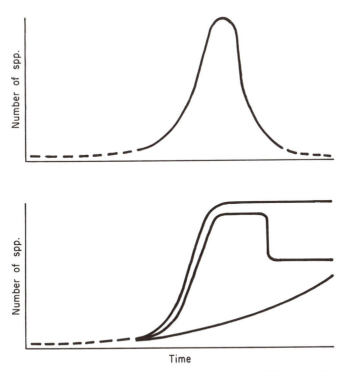

Figure 18 Pattern of change in diversity within the temporal history of a clade: (A) model curve showing the situation observed for the most groups of animals; (B) Curves that are not observed in the fossil record.

diversification of groups tend to be extremely similar (Figure 21). The reasons for this phenomenon are not clear. They surely have their origin in the hypothesis of the Red Queen (van Valen, 1973, see 1985), according to which the effective environmental world is constantly deteriorating. A group eventually cannot run fast enough to keep up with changes either in the physical world or the competing biological world. All is change; decay inevitably accompanies success. A complementary explanation may be that, within a diversification, taxa tend to become more and more specialized as they divide up between more and more numbers of taxa an essentially limited domain—the domain defined by what that group does well. Increased specialization inevitably leads to increased vulnerability to any perturbation of the system—physical or biological.

Certain members of some groups may persist for very long times, after the principal diversification of the group has been long forgotten. These are the living fossils. Darwin coined the term and used living fossils as a test of his theory of natural selection. If natural selection is a cause of evolutionary change, then the effects produced by the mechanism should be in proportion to the intensity of the causing process. Darwin thought that in certain environmental circumstances, such as freshwater basins, "competition . . . will have been less severe than elsewhere; new forms will have been more slowly formed and old forms more slowly exterminated" (Darwin, 1859). Today there are multiple possible explanations of the persistence of living fossils, most of which probably have to do with historical contingencies rather than the applications of any single principle (Schopf, 1984; Eldredge and Stanley, 1984; Thomson, 1987b).

Finally, the success of the diversification of a group seems to be something to be measured differently at different focal levels. For the individual and the population it may be in terms of survival and fecundity; for higher taxa it is in terms of taxonomic diversification and persistence. For all levels it is to be measured in terms of relative efficiency at utilizing energy.

The relationship of adaptive radiations to the origin of major groups is a particularly interesting and difficult question. As a broad generalization we can say that the course of evolution tends to go from less to more specialized. Occasionally, however, the pattern of diversification of a group at lower taxonomic levels is punctuated by the appearance of a new group that redefines the terms upon which diversity is built, allowing a whole new range of adaptive options. We then call this a new major group. For example, mammals are strictly only a subgroup of the mammal-like reptiles, which are in turn a subgroup of the reptiles, which are a subgroup of the tetrapods, and so on. In terms of tetrapod characters they are rather specialized, but mammal characters form the primitive basis for a huge set of radiations of diversity. Systematists are fond of pointing out that if we only knew *Archaeopteryx* and a couple of Early Jurassic mammals, we would have no hesitation in classifying them merely as aberrant reptilian forms. The definition of major groups includes a great deal of hindsight; nonetheless, we must ask what it is that allows the triggering of new diversification on these scales during the history of the group. The answer is that the process must in large part be historically contingent, because diversification is both an intrinsically and extrinsically controlled and driven process. Our aim here is to look at the intrinsic part of this, and immediately it becomes clear that when a new major group is defined, it is always based on a number of fundamental reorganizations of morphological systems, rather than the accumulation of large numbers of the sorts of trivial character differences that produce species-level diversifications. The crucial points in the history of any group are those that add a real level of complexity, as opposed to allowing a degree of further diverse specialization.

THEME AND VARIATION

All of the above leads us to ask questions about what the general properties of evolutionary diversification are; it is essential that we resolve this problem if we are to discuss how developmental processes contribute to their cause. I will argue from two very straightforward propositions.

First, evolution is all about making differences between and among entities. As a process, its end result is to produce gaps. The basic feature of all diversification, at whatever level, is therefore the reality and the importance of gaps and differences. These in turn are only recognizable in reference to the background of continuity and sameness that is broken by the evolutionary process. That all groups are separated by gaps is obvious at higher taxonomic levels, but is equally true at the lower levels: species, populations, individuals. Even where species or populations appear to grade insensibly into their nearest neighbors in space or time, nonetheless there are differences to be recognized (or else we could not have distinguished those neighbors). Obviously some characteristics show continuous variation (size, for example, often does), but equally obviously many do not. Explanation of the latter is particularly important.

The second proposition follows directly from the first. It is that the distribution of organisms within any diverse group (individual within populations, populations within species, species within higher categories) is always clustered. That is to say, there is a hierarchy of gaps. At each level, we find groupings of related forms with appropriately scaled gaps between them. There may be several subgroups within a given level (species within a genus, families within an order). But the gaps between groups are always bigger than the gaps within groups. The processes of evolutionary diversification therefore have produced clusters of more closely related forms: clusters of clusters in fact.

The patterns of organismic diversity predicted by classical evolutionary mechanisms should form a seamless web. Or perhaps diversification proceeds like a slowly flowing three-dimensional river. In such a continuum there are few if any gaps; there are, of course, spaces among the major and minor flows, but not within. Within groups, change should be continuous, smooth, and basically gradual. As Darwin wrote in his "Essay" of 1844, "as I suppose that species have been formed in an analogous manner with the varieties of domesticated animals and plants, so must there have existed intermediate forms between all the species of the same group, not differing more than recognised varieties differ." In such a model taxonomic boundaries are artificial, products of a human need to categorize and of the unavoidable deficiencies of the record. The essential continuity of the process ideally would be expressed by continuity of both morphology and systematics. All the required intermediate forms once existed, if only we could find them. This is a pretty good model, in fact. It certainly follows logically from available theory. Obviously there have been attempts to modify the model, one of which is Eldredge and Gould's punctuational model (Gould and Eldredge, 1977) of speciation, which depends on Mayr's model of peripheral isolates and genetic revolution.

But theory ought also to derive from data. In this case the data are the observable patterns of organismal diversity. Do patterns of organismic diversification really support such a continuous scheme? I suggest that they do not, and that a different model more accurately describes the pattern of evolutionary change over time and in space. The model I propose does not require any radically new evolutionary mechanism. It is merely a different way of looking at the same old problem. It is a model that focuses attention on the gaps in and among diversifications and on the nature of diversification itself.

Most evolutionists have had a little trouble in dealing with gaps in the record. Darwin surely provided us with most of the explanation in pointing out that, to take the simple example of species A giving rise by splitting and divergence to species B and C, we would not expect to find intermediates between B and C, only intermediates between A and B and A and C. Those intermediate stages were, by definition, bound to be less successful than B or C, and the fossil record in unlikely to have picked them up. The inevitable corollary of this sort of explanation is that some species or populations are much more successful than others. But this is not the end of the matter. If B and C now split and species D, E, F, and G are produced, some of them should in theory be more successful than B and C and evolution will continue—a slowly flowing river (a clade) of changing species all descended from A.

I would like to suggest, however, that the gaps in the fossil and Recent record, although they are traceable as gaps in lines of descendancy (a species- and population-level phenomenon), also stem from the nature of diversification at the species level and above (a macroevolutionary phenomenon). I suggest that the course of evolution is modeled better as a series of clusters of species than as lines of species. If one looks at any reference work on systematics of a group, one finds immediately that the diversification of groups does not demonstrate within-group progressive evolution (a stream of slow and continuous change) but rather a set of equivalent variations around a fixed theme. In the *Aves*, for example, all woodpeckers are remarkably alike, as are all hummingbirds or titmice. This is, I submit, not just because intermediates between, say, woodpeckers and their nearest sister group have been lost from the record (if they ever existed) and that woodpeckers have not yet given rise to anything new, or even that woodpeckers have given rise to something new that we recognize as part of a different clade (which could also be true, given the logic of taxonomy). It is also the case that woodpeckers or titmice only give rise to other woodpeckers or titmice. The vast majority of evolutionary change is not progressive but conservative, a form of "running-in-place." The record appears to have gaps because evolution principally produces clusters of the same thing. This is a macroevolutionary pattern of major importance.

I suggest therefore that the principle mode of evolutionary change, as demonstrated by the living and fossil record, is not open-ended progression (however slow and gradual) but sets of clusters that we can call "themes and variations." "Theme" means different ways of doing things biologically and variation here means the range of phenotypic expression of organisms based on that theme. For example, the two groups of populations that we define as the hairy and downy woodpeckers (*Picoides villosus* and *P. pubescens*) are but two sets of variations on a theme that we name the genus *Picoides*, part of the broader theme we call woodpecker (Picidae). In relation to systematics and taxonomy, "theme" here therefore roughly corresponds to "higher group" and "variation" to "included group." All woodpeckers are variations on a basic theme set in terms of a particular morphology including the beak and specialized tongue and foot structure. (We must always remember, however, that the systematics of any group is still an imperfect and artificial construct.)

The notion that phenotypic variation is not normally unlimited and open-ended has already been expressed at the population level by Oster and Alberch (1982), who argue that developmental constraints will always operate so as to limit the number of nodal positions in morphotype space that can actually be realized. I would argue that the evidence of organismal diversity in space and time demonstrates that a similar phenomenon must operate at all levels. The limitation of phenotypic variation is not the operation of selection (and the time over which it operates), but rather the properties of a given morphogenetic system. For any given phenotypic feature (and thus for any defined taxonomic level) the intrinsic morphogenetic system allows both a given range of immediate phenotypic expression and a given fixed range of potential future variation. This envelope of potential variation is normally distinctly bounded, the bounds being intrinsically determined by the limitations to possible disturbance of the developing system. Mechanisms like selection simply have to work with these givens.

At the level of defining species or lower levels, essentially all the range of phenotypic expression will consist of variation in characters that are caused at a morphogenetically late or superficial level (see preceding chapter). These may involve the cytodifferentiation stage or late pattern expression in relatively minor characters. The limits on such phenotypic expression are set both by factors operating at the developmental level and (importantly) by factors operating at more fundamental levels. At taxonomic levels above the species, the defining and enabling characteristics are caused at deeper developmental levels involving more fundamental aspects of pattern formation in morphogenesis. In the example of a family of birds like the woodpeckers (Picedae), within-genus species-level morphological differences (we will not deal here with behavioral and ecological factors) are all in matters of plumage color and pattern, size and minor differences in proportions. The same is seen in other bird families.

At the family and ordinal levels different characters are involved, reflected in more drastic differences in body proportions and adaptively related characters such as feeding and locomotor mechanisms. In general, high level characters involve modes of doing different things while lower level characters involve different ways of doing the same thing.

A key element of the theme and variation model I propose is that the range of possible variation expressed within an entity—say, demes within a single species, or species within a "genus"—is fixed. Or rather, it is limited rather than open-ended. Now obviously this must be so within any short time frame. I propose that it is also effectively fixed over longer time frames. In order to change the potential range of phenotypic expression it is necessary to change the underlying mechanism. Further, for each level of control over phenotypic expression there is a different and potentially discrete causal mechanism.

I propose therefore that effectively, for any biological level in the hierarchy, there is an intrinsic "theme" and a limited set of variations. The limitations are presumably due to developmental constraints and other internal factors that are shared by all individuals within that unit and that operate at every level from the earliest genetic to the latest morphogenetic. That such limitations operate at the species level is possibly indicated by the well-known fact that strong artificial directional selection eventually becomes ineffective (although there are problems in extrapolating from artificial selection to natural selection, as discussed in the next chapter).

This is a model that seeks to explain the relationships between individual taxa (variations) within an including higher taxon (the theme). The included taxa (say, species within a genus) all share the common characteristics defining the including taxon in systematic terms. They also share features of developmental systems. I suggest that the developmental constraints that they share define the themes and the possible ranges of phenotypic expression. In attempting to visualize this model, I find it convenient to think of the phenotypes as points on the surface of a hemispherical plane or, better, a bowl with a flattened base (see Figure 19). The surface area defines the potential range of characteristics. Given points on the surface represent realized or realizable sets of characteristics—a species or higher taxon. Evolutionary change and diversification within a taxonomic unit is represented by a ball moving on the surface and stopping at certain points. The bowl-shape of the surface models a situation in which certain variants (those near the center of the surface and on the flatter portion) are easier to realize. The fact that the ball can stop at certain points but not others reflects the operation of developmental constraints and morphological integration (and thus the surface has properties of modeling an adaptive landscape). As one gets further and further from the center, more energy is needed to move the ball (evolutionary change becomes more difficult). The amount of realized morphological change (lateral distance moved) per unit of "enery" used to move the ball becomes less, or to move the ball a given lateral distance more and more "energy" is required. The steep vertical wall of the rim represents "forbidden" territory. It should be noted that in this model, not

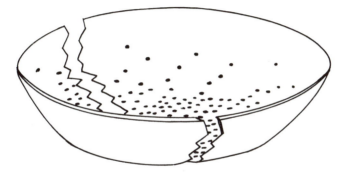

Figure 19 The bowl model. Dots represent realized or potential positions within the curved morphospace.

all potential points need be realized. The existence of one included taxon is enough to express the nature of the underlying theme causing the shape of the envelope of allowable phenotypic expression.

In this model, the taxa lying near the rim of the surface are more difficult to make; to get to this position requires more "energy," and even more is required to propel the ball over the rim onto a new surface. What are the biological equivalents of "difficulty" and "energy"? The most obvious factor involved is change in morphology, behavior, or other characters affecting adaptive characters (fundamentally characters affecting the use of metabolic energy—van Valen, 1976). It is apparently difficult to break an allometric relationship, for example. Thus movement to a peripheral position defined by "increase in size" may be difficult beyond a certain range because of concomitant mophological changes (the Irish elk effect). Characters must be functionally correlated. It is no use changing the proportions of the limbs for greater speed if the metabolic system cannot be changed at the same time to provide the energy to move the limbs. Morphogenesis of characters must be developmentally correlated. Strong selection for change in a particular characteristic may result in a disturbance of the balance of development of other characteristics. This is developmental constraint. Again, certain changes can be affected by quite minor genetic changes while others are only possible with major genetic changes. Major change is potentially disruptive. Within each "theme" level, therefore, there exists a sort of equilibrium situation where the evolutionary changes that are produced are principally balanced and coherent and there are strong intrinsic reasons against the production of oddities.

Below the species level, the model operates somewhat differently because populations potentially interact, and through gene flow there is the possibility of actually changing the points at which the ball can come to rest.

If the theme and variation model is correct, it must follow that the vast majority of all species ever produced have little or no potential for progressive evolution. They are created as minor variations within a single late morphogenetic or cytodifferentiational process. What Gertrude Stein said was correct:

"A rose is a rose is a rose." It is on this basis of equivalent variations on a single theme that most diversification within a group proceeds. But some species must have a greater evolutionary potential. They are the species that break the pattern and may (if other circumstances, mainly extrinsic, are correct) form the basis of a new pattern and thus a progressive evolutionary change. Such species can only be those that differ from their sisters by virtue of changes in the deeper, more fundamental elements of patterning systems. In this case, the immediate change in phenotype of the new species may involve only a very subtle difference at first; large saltatory leaps need not be involved. Unless there is some linked effect, in terms of those characters by which their sister species vary, these "key" species may appear to be a normal part of the cluster sisters, because that particular set of systems has not necessarily been changed. However, another crucial set of features has changed, and it is on the basis of this that the limits of the original theme may eventually be transgressed and a new theme started.

In this model, all species within the theme are equivalent. Therefore, in practice the immediate genealogical sister species of any other species in the cluster is not necessarily that which occupies the closest adjoining part of the morphospace. Within the theme any species could be the sister of any other. If correct (even if the occurrence is rare), this makes it extremely difficult to reconstruct phylogeny. Any cladogram of the species is likely to show lots of confusing convergence, homoplasy, or even apparent reversals. [This may be the case in the labroid fish radiation analyzed by Stiassny and Jensen (1987).]

Perhaps the best example of "theme and variation" is that of mammalian origins, considered above. The character changes involved in the origin of mammals were a series of transformations, most quite subtle, of such basic adaptive characteristics as the jaw mechanisms, locomotor systems and posture, and sensory systems and brain. Quantitatively, the differences involved between levels were quite small. At each "stage" in the process, a new species-level diversification of organisms was allowable, based on the subtly different ways of being a mammal-like reptile. These radiations must first have involved species-level characters of the normal sort, such as color, odor, and size, by which the ancestral group diversified at the species level. Eventually, through the sorts of feedback mentioned in the next chapter, the new biological patterns will have allowed different sets of species-level characters to be important. Thus in the end, mammal species distinguish each other by, and compete with each other in terms of, different types of characters from those reptiles use. They are not, however, characters of the jaw joint or gait.

This sort of model produces a fundamental discipline on the way we use comparative biology. Comparisons of taxa that belong to the same hierarchical level (are equivalent variations on a given theme) will be less useful, for example in trying to relate phylogeny and ontogeny or in discovering atavisms, than comparisons between levels.

In the next chapter I will argue that the sort of radical or quantum change (to use Simpson's word for the same thing) causing between-level (as opposed to within-level) novelties is likely to be literally (as well as figuratively) a threshold effect. Here I simply propose that the pattern of evolutionary diversification that we actually see in the living and fossil record is one that fits a theme and variation pattern better than a continuous seamless web. The essential feature of what drives this clustering process is the control of different levels of phenotypic character at different levels of morphogenetic pattern control in development.

How can we test the model? It is important to see that the theme and variation model allows gaps and predicts them, but does not require them in all cases. It is most likely that in most cases the range of variation of two adjacent themes will be overlapping. Similarly, the theme and variation model does not require that all potential variations actually be realized. In most cases most variations will never occur, due to chance. And still many others that have occured will in fact not be found in the record. In principle the theme and variation model should become more and more visible at higher and higher taxonomic levels. Indeed, as we have noted, at the level of individuals and populations within a single species it is very similar to current "continuum models." It applies most strongly at the level of speciation and above and thus is concerned principally with the phenomena of so-called macroevolution.

If the model applies, it should be the case that within a given theme there will be higher rates of taxonomic evolutionary change and lower rates of morphological change. At the point at which a morphogenetic threshold is crossed there will be higher morphological rates and lower taxonomic rates. In the end the view I propose here will be tested by the extent to which it is explainable in terms of developmental mechanisms.

Finally, a note must be added concerning systematic characters. In the preceding discussion we have been concerned with the evolutionary development of characteristics that produce new taxa. In the case of higher taxa, these are not necessarily those by which systematists now diagnose the same groups. The characters on which the systematics and working taxonomy of higher groups are based tend to be a mixed bag. The list of available shared derived characters for a major group tends to be a list simply of what happened to be left in common after millions of years of diversification. In fact it is only the subset of those characters that happens to have been spotted by the taxonomist. The characters of higher groups that working systematists list in their monographs are therefore not necessarily the full suite of those characters that "caused" the origin and allowed the radiation of the group.

9

Morphogenesis
and Evolution

In the preceding chapters we have built up the premises of an argument concerning morphogenesis and evolution. These are as follows:

1. Evolution occurs through processes of introduction and sorting of variation.
2. In a genealogical hierarchy, introduction and sorting of variation can occur at a series of focal levels and it is a property of the hierarchy that there is upward and downward causation between focal levels.
3. The crucial role of developmental processes with respect to evolutionary mechanisms is in the causation of new phenotypes. Phenotypes are always expressed in individual organisms, but the properties of other focal levels in the genealogical hierarchy must also be considered.
4. Development is also essentially hierarchical, involving processes acting at different focal levels. Each level is defined as the place where new gene expression occurs. For simplicity we can divide its hierarchy into stages from early pattern formation to late cytodifferentiation, each including unknown (but very large) numbers of phases of new gene expression. It must be noted that ontogeny of a given individual or given taxon represents a route through the basic hierarchy of developmental stages that, through historical accident or selective bias, may be extremely convoluted and unpredictable.
5. In the systematics of any group, there is a general correlation between taxonomic rank and different grades or ranks of morphological characters. There is a series of levels or grades of generality of phenotypic characters caused at different levels of the morphogenetic hierarchy.
6. The morphogenetic hierarchy that produces the different grades of phenotypic morphology can potentially involve upward but little if any downward causation.
7. The course of evolution appears principally to produce clusters of evolutionarily equivalent species rather than lines of progressive change. At any taxonomic level, diversification within a group has therefore to be distinguished from those rarer phases of progressive evolution leading to establishment of new groups. The two involve quite different processes because they involve the morphogenetic causation of different levels of morphological characters.

120

8. Phenotypic characters may occur in groups that are linked both in the sense of functional integration of the phenotype itself and/or by virtue of the integration and interdependency of developmental pathways in morphogenesis.

Given these premises, we can explore a whole range of questions based on the central problem: how do morphogenetic mechanisms change and what general properties of evolutionary mechanisms are affected by the general properties of such systems?

The basic question, which lies behind most attempts to explore the relationship between development and evolution, is: how are large-scale reorganizations of the phenotype over evolutionary time caused? We should set aside the question of rate of change for the moment. We need to concentrate first on the scale of change. While something like a change in eye color in *Drosophila* from red to white is surely a major change, when measured on the appropriate scale, large-scale usually means a change in a character complex of major taxonomic importance. An example would be the origin of the dentary-squamosal jaw joint in mammals, or the evolution of oligosyndactyly in horses.

There are two possible types of mechanisms. First, large-scale reorganizations might occur as the result of accumulation of smaller scale phenotypic changes caused solely by late-stage morphogenetic processes. Second, large-scale phenomena might be caused as the consequence of changes in earlier stage morphogenetic processes that affect major levels of phenotypic characters control directly—particularly those stages where more general aspects of morphological pattern are controlled. In the latter case, we have the possibility in theory at least that large-scale changes are caused through very small initial changes that allow disproportionately large results to accumulate slowly through gradual evolutionary processes by fixation of new alleles throughout the developmental cascade. Alternatively, the initial change might produce a revolution in phenotypic structure by an immediate reshaping of the developmental cascade. Whatever mode operates, modification of the phenotype requires a reprogramming of morphogenesis.

There is, however, a very widely held notion that evolutionary changes cannot be caused by changes in the deeper, more fundamental events of development. Such changes are often thought inevitably to be lethal. After all, so the argument goes, very many variants that seem to be caused at relatively late stages are lethal; safely changing more fundamental stages must be even more risky. The whole notion of a "macromutation" that would significantly alter the early course of development of the phenotype is, in this view, fanciful. It must follow, therefore, that all evolutionary change must occur in terms of variation introduced at late developmental stages and in the form of very small-scale changes in phenotype. By no coincidence, these are the sorts of changes that are amenable to study in quantitative and populational genetics.

Despite these traditional skeptical views, it seems inescapable that changes in early developmental systems are possible and we have to find out how they occur, rather than attempt to dismiss them. First, we should note that a changed early developmental event need not be a "macromutation" (the term is pejorative) in the sense of producing a major change in phenotype immediately. Second, we ought also to admit that it would be very difficult technically to recognize a major revolutionary shift (should one have occurred) because we are conditioned to view any gap in the living or fossil record as an artifact rather than real, and are conditioned to assume that all changes are slow and smoothly continuous. Eldredge and Gould's concept of punctuated equilibria at the population/species level is only just beginning to break this habit of thought and observation. But perhaps the most telling evidence that early developmental systems must be modifiable is so obvious as to be almost laughable. For the fact is that, if one compares morphogenetic patterns among organisms, they obviously *do* differ at all stages, from early to late. This is again a phenomenon where scale is important. We all know that at the earliest stages of development different taxa look alike. But, as von Baer's first and second laws clearly show, early stages only look alike in a progressively more general frame of reference. A frog and chick embryo, at various stages after gastrulation, look alike as chordates, vertebrates, and then as tetrapods. At each stage of development they also differ. The observable differences at any early stage may seem infinitesimal to us, but they are obviously crucial to the embryo concerned, because they represent the foundation for the subsequent divergence of development.

The apparent conservatism of early development is thus partially an artifact. It obviously stems from the fact that the processes of early development are extremely similar in related organisms. But they are not identical. They appear similar to us because we have not yet discovered all the subtleties by which they differ at the extremely fine scale that is appropriate to that level of developmental complexity. But there are obvious clues in, for example, the different early development of amphibians with respect to sequestration of presumptive germ cell material. We need not equate the similarity of the morphological appearance of early embryonic stages with identity of process (particularly at the gene level) or with impossibility of change. If organisms are related, divergence of developmental pathways must be possible at every stage from gametogenesis to late cytodifferentiation.

Such divergence can only have occurred in one of three ways.

1. It may have occurred during the course of significant change (especially an increase) of developmental and morphogenetic complexity. That is to say, from an ancestral state, whole new developmental stages were added, and added differently in different lines. This was surely the cause of the divergence of phyla of animals and plants during the first origins of multicellularity. In each line, the course of development after the point of divergence must have been significantly different right from the beginning of the change. Further, it could occur through reprogramming of a given set of developmental pathways (potentially totally independent of change in complexity). This must be the case in, for example, the origin of new families of insects. The second mode has two alternatives: our second and third possibilities.

2. One mode is that of accumulation of phenotypic differences caused at the late stages of morphogenesis and the subsequent reprogramming of morphogenesis at earlier stages to accommodate these later changes (this is a type of downward causation).
3. The last possibility is the causation of change at earlier stages with concomitant effects at later stages (a form of upward causation).

In all cases, one must never lose sight of the fact that any change has to be accomplished through the divergence of populations and species.

INCREASE IN COMPLEXITY

Evolutionary divergence through differential increase in complexity of morphogenetic systems is obviously an essential element in the explanation of the origin of different major groups of organisms. When we come to lower taxonomic levels (e.g., species or genus), it is hard to find evidence of the origin of new cell or tissue types or totally new types of morphogenetic processes. The whole history of the vertebrates, in fact, is extremely consistent: very few totally new features appear. What seem to be new (e.g., the mammalian ear ossicles) turn out usually to be old established structures transformed (part of the jaw suspension in this case). What we find instead is differential modulation of a single common system especially at lower versus higher taxonomic levels.

One place that offers an especially interesting opportunity for change in complexity, even in well-established groups of organisms, is of course the whole field of life history phenomena. There are particularly good possibilities in groups that have complex larval life cycles and metamorphoses, and this is no doubt a major factor in the evolutionary diversification of the Insecta. Geist (1978) has made some extremely interesting observations on life histories in mammals in this vein also.

While the most common sort of change in any developmental system is probably the causation of quantitatively different phenomena (e.g., rate of cell division), increase in complexity must eventually involve qualitative differences. At its simplest, an increase in complexity of developmental information might start in one of two ways (although the distinction is a bit arbitrary). Either a signal given at a particular decision point in a cascade might change or the responding system might be modified. In either case, the result must be that a cell population normally responding to a particular signal in a particular environment in a unique way has to fail to respond as a whole. A bifurcation of development occurs when part of the responding tissue receives the signal, or when part responds differently to the signal. From this point, while part of the system functions normally, a subset has been sequestered that has behaved differently. But both continue potentially to be subject to all the late events of the cascade. Any such change therefore has consequences all down the line. There is now a subpopulation of cells that does not fit the following series of controls of development. One can imagine the predecessor of the vertebrate neural crest as having arisen through such a missed signal in the induction of a lateral strip of neurectoderm in some primitive chordate. The new population struggled along responding progressively differently to signals received down the line, torn between conflicting neural and mesodermal influences.

In order for a simple bifurcation to occur there must at minimum be a simple quantitative change in either signal or responding tissue. In a gradient system, for example, if the signal were normal in type but weak in strength, distant parts of the normal responding system might not receive it, or might receive a sufficiently weak dose as to respond abnormally. Equally, if the responding system were of different volume than normal due to change in the pattern of prior cell division, then parts of the responding system would receive abnormal signal strengths from a normally functioning sender. Similarly, in the case of tissues that change position in the embryo, a changed signal–response system might arise from changed timing of interaction. Another type of bifurcation might occur when a normal population of migrating cells encounters a mutated set of ECMs. Changes in ECMs might cause a mismatch of part of the combination by changing the timing of relative movements, and vice versa.

Obviously it is not always appropriate to use the terms quantitative and qualitative as polar opposites in these considerations. Any change caused as a shift in a quantitative parameter is liable to have qualitative consequences further down the morphogenetic cascade—and vice versa. For example, a very subtle quantitative change in the course of rearrangements occurring in vertebrate gastrulation might have a qualitative effect in terms of the inductive relationships occurring between chordamesoderm and neurectoderm. The prechordal mesoderm and its inductive role with respect to the vertebrate forebrain may have arisen in early vertebrates by means of such a simple initial change.

In general, complexity of morphogenetic systems and consequent phenotypic morphology correlates with the levels of taxonomic diversity produced. Given what we know of phenotypic morphology, we have to assume that change in complexity (usually, but not always, increase in complexity) was important in the origin of higher taxonomic categories and was progressively less important in the subsequent diversification of those taxa, which is to say, in the origin of lower level taxa. Diversification at the species level usually does not involve change in complexity of the phenotypic morphology or the generating morphogenetic system. This raises the question of whether any change in complexity (particularly an increase) automatically renders any further change less likely until, in what we would consider complex organisms, further significant change in complexity becomes virtually impossible and one is left only with diversification based on modulation of the established pattern. This brings us to the subject of reprogramming of development.

REPROGRAMMING OF DEVELOPMENT

In the hierarchical morphogenetic cascades that make up developmental pathways, variation can be introduced at any stage where new gene expression is caused. What happens to that variation: where is it sorted? Obviously any variation that causes variation in the final phenotype will be sorted at the level of individual organism in an ecological context. But before this can happen, developmental variation will be sorted within the context of development itself.

As discussed in Chapter 7, the cascade of gene expressions involved in developmental pathways involves an alternation of phases. A given phase of gene expression lays down new boundary conditions for cell and tissue-level processes. These processes will involve intrinsic rules and discrete initial conditions. Under the influence of some sort of instructive signaling new gene expression is caused, and the cycle starts again. As a result, the behavior and nature of the sets of cells and tissues concerned become progressively transformed (usually highly interactively). Any part of this system is liable to change, principally from variation in gene expression.

New gene expression within a pathway will be sorted according to its compatibility with the existing environment of the rest of the pathway. New gene expression in interacting pathways will be sorted according to compatibility with expression in the first pathway. If there is a change in the signaling system acting on a pathway, the result could be a changed response on the same scale as the change in signal or on a very different scale. However, probably only rarely will there be a simple linear relationship between the scales of cause and effect. Most changes in gene expression within pathways will probably simply be accommodated, without affecting the phenotype. Then one more change may be added that forces the system over a threshold into a new mode, even though (especially earlier in development) the change produced is extremely subtle.

Downward Causation

We have already discussed the fact that different levels of phenotypic character must be caused at different morphogenetic levels (Chapter 7). Whether there is any downward causation in morphogenetic systems is an interesting question. In general the answer appears to be no. If we consider once again the example of the morphogenesis of skeletal patterns in the vertebrate limb, the reasons become obvious. The basic pattern of elements is laid down very early. In terms of phenomena observable with the eye, it occurs when the precartilaginous blastemata are formed. In fact, it occurs before this in the activities of the AER and ZPA relative to the undifferentiated limb mesenchyme. A portion of the control over size and fine details of structure of the skeletal elements is effected relatively late in terms of gene expression acting at the chondrogenic and later stages. It is hard to see how genetic variation introduced at this level could affect the basic pattern laid down earlier, not just because of the relative timing but because the later batteries of gene expression involve fundamentally different developmental processes. Selection on those later stages that involve the genetic control of chondrogenesis or osteogenesis will not affect the gene arrays that control the earlier stages of pattern formation. They could not sum up to effect the earlier phases of morphogenesis.

True downward causation would involve a direct cause-and-effect relationship between phenomena at later and earlier stages in morphogenesis. An indirect effect of great power, however, would arise from a feedback between phenotypic character states. As a model, we can examine three separate stages in morphogenesis of a mammal: late expression of coloration in the phenotype, late pattern expression of size and proportion, and early pattern expression of discrete musculoskeletal configuration. Let us imagine that the color alternatives possible are dark versus light, and that the musculoskeletal configurations involved are in limb architecture. The three sets of phenotypic characteristics controlled at these three morphogenetic levels are bound to be integrated in different ways with a huge number of other phenotypic characteristics and their control. Let us concentrate on the fact that they are all involved in some way with matters of metabolic energetics. In a terrestrial mammal (or a lizard or bird), coloration will affect heat loss and gain, size and proportion will reflect total energy budgets, and particular musculoskeletal configurations will affect the efficiency of prey capture. Feedback among these components could potentially occur in several ways (Figure 20). Of evolutionary significance is the fact that there will be a feedback in "permission" (rather than causation) that affects fixation of alleles with different morphogenetic consequences. One feedback loop would be that darker color yields a change in the energetic regime such that body size change and eventually change in limb proportions would be favored. Such a positive feedback loop would be open-ended until some limiting factor was reached. Going the other way, changes occurring first in the fixation of new limb proportions might correlate with higher energy intake and thus favor change in coloration patterns that reinforce a different metabolic budget. Most feedbacks would probably be negative or self-limiting, but where the feedback was positive, it could be an immensely powerful driving

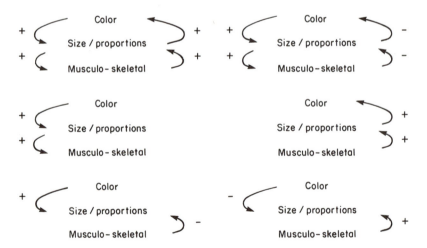

Figure 20 Some possible interactions in a feedback system comprising three elements: color, size and body proportions, and musculo-skeletal configuration. Arrows represent direction of causal influence; + signifies positive feedback; − signifies negative feedback.

force for an internal sorting of genetic variation, both at the same and different levels, biasing that sorting in particular directions. Here externalist and internalist mechanisms combine in a striking way.

This sort of "morphogenetic feedback mechanism" is at the very heart of the pattern of "correlated progression" discussed in the previous chapter. It accounts not only for the linkage of character change in a time sequence, but also the scale and speed of those changes in many cases like the origin of tetrapods or of mammals. Once again, there will be a difference in effect between feedback among processes acting at the same level and those acting at different levels. The former is most likely to give graded, small-scale effects. The latter has the possibility of producing larger scale effects as well.

There might also be a range of indirect effects involving downward causation. For example, if selection on later stages involved inversions or other gene rearrangements, these could have side effects that affected neighboring genes that are expressed at much earlier stages. As noted earlier (Chapter 6), there is likely to be fairly strong selection for "housekeeping" changes whenever there is loss or reduction of a morphological character. We have already noted that if morphogenesis of an element of the vertebrate limb were to fail as a result of a mutant acting at the ossification stage, it would continue to be formed at all the prior stages, because the blastema stage is essentially invisible to selection acting on variation in genes expressed at the ossification stage. However, once an element were lost at the ossification stage there would be an internal housekeeping "incentive" for reprogramming of morphogenesis by fixation of any alleles that reduce the investment in earlier stages. This phenomenon would be strongly limited by the interactive nature of development. The early stages do not operate in isolation but as part of a whole. The earlier stages of any element are an intrinsic part of the morphogenesis of other elements.

In summary, direct downward causation in morphogenetic systems seems unlikely, but a certain range of feedback favoring the fixation of genetic variations reprogramming morphogenesis may be possible.

Upward Causation

Given the cascading nature of morphogenetic pathways, upward causation of effect from changes introduced at earlier stages to later ones seems not only likely but, in many cases, inevitable. In terms of the model of pattern control outlined in Chapter 7, each stage of morphogenesis depends on the correct convergence of conditions created by preceding levels in the cascade (new phenomena with their boundary and initial conditions) and the controlling signals and conditions for the new stage. Any change in a morphogenetic pathway will potentially have both immediate and long-term ramifications. The highly buffered and canalized nature of devlopmental pathways and their mutual interdependency make it likely that many changes introduced at a particular level (through genetic or other aberration) will be corrected or compensated for (to an unchanged phenotype), especially if the changes wrought in the system are initially quantitative—cell number, for example. In a longer time frame, changes introduced into the system may also be corrected by fixation of alleles that canalize the system for the original pattern. But for any system there will be a threshold level, beyond which change introduced at any level in the system will inevitably be passed on permanently to the subsequent stages and will create the possibility for fixation of further changes. The accumulation of "minor" canalized changes will inevitably bring the system to a threshold where one last straw will trigger some significant change.

Here it is worth emphasizing that many changes produced in earlier morphogenetic stages are indeed likely to take the form of quantitative variation, perhaps mostly in the initial conditions of any stage. It is here that the most capacity for correction of earlier changes is possible, through the quantitative aspects of subsequent stages. But once qualitative differences between stages are involved, correction may not be possible. In the limb, blastemal growth phenomena might correct initial deficiencies in mesenchyme supply. Even remodeling processes in ossified elements may affect size of bones. But once a blastema has branched in growth (or failed to), then an irreversible "decision" has been made that will be unaffected by later stages. The ossification stage, for example, will simply ossify what has been presented to it.

Once a change has been caused at any level, and its consequences felt at other levels, the system is disrupted. This disruption will have two complementary forms. First, it will potentially change the integrity of the developmental pathways involved. Second, the resultant phenotype will have a new form and thus a new status in terms of potential functional parameters. This is the basic stuff of any evolutionary change. We are simply accustomed only to think of the phenomena in terms of very simple late orphogenetic expression of essentially minor phenotypic characters (i.e., the data base of population biology). We must allow that the same properties pertain to characteristics caused at early morphogenetic stages, whether major or minor.

We should expect that there will be sorting of any changes occurring in developmental pathways in terms of these two sets of "integrities." Genetic change at any stage of development is sorted in terms of its fitness with respect to the genetic and morphogenetic environments in which it occurs and is expressed. If the resultant phenotype is altered, there will be a reinforcement of the genetic/morphogenetic sorting processes through a strong feedback loop.

As noted with respect to downward causation, a feedback in the sorting of variation at different levels of phenotypic character can act as a powerful driving force of change here. Because different levels of phenotypic character are caused, or at least initiated, at different levels of the morphogenetic hierarchy, there is a potential for the setting up of complex sets of feedback between processes affecting the sorting of phenotypes and the sorting and fixation of genetic variants in the morphogenetic pathways (Figure 20). This will be reinforced by the internal processes of upward causation in the morphogenetic pathways.

However, a very important phenomenon now takes over. Whereas some phenotypic characters (those caused late in morphogenesis) show more-or-less continuous variation, earlier developmental systems are not likely to be so smoothly changeable. As Kauffman (e.g., 1983) and many others have noted, it is a property of complex interactive (constrained) systems that, when subjected to stress they respond, both immediately and in terms of long-term evolutionary change, either by no change at all or change to a discretely different "nodal" condition (threshold effects). The scale of change may be quite small, but the pattern is one of shift from one stable pattern to another. I submit that this is the basis of clustering effects in systematics of organisms. This behavior will accentuate the capacity for both stasis and change in phenotypes and militates against the production of certain intermediate phenotypic conditions that are adaptively possible but developmentally prohibited.

Thus we can see that at one extreme the consequence of change in developmental systems at any stage, through upward causation, might under certain circumstances have little or no effect on the phenotype. This could be because the change actually was of very small scope, or because the potential range of change was corrected within subsequent stages. But equally, a small initial change at some level in the morphogenetic cascade potentially could have a more significant phenotypic effect through threshold change, moving the morphogenetic pattern to a new "nodal position." A prime example of this would be the case of the origin of different patterns of the reptilian ankle joint (Figure 8). Here the possible phenotypic conditions represent a series of binary choices, having to do with the relative arrangements of limb rudiments and muscle rudiments. Some potential intermediate phenotypes are impossible developmentally, while others could be expressed but would be immediately strongly selected against. The cause of the variant conditions is to be found in the early morphogenetic stage of limb development. Similarly, reorganization of the limb in horse evolution may involve the repatterning of the mesenchymal blastema at a very early stage. In the case of horse evolution, many intermediate phenotypic conditions, represented by partial (as opposed to complete) loss of lateral digits, may never have occurred or may have been ephemeral in the extreme, because of threshold morphogenetic phenomena in the reprogramming of pattern control in the digital arcade, reinforced by very strong selection against any intermediate phenotypes that made it through the system.

According to this view of morphogenesis, change in morphogenetic pathways can be initiated at any stage and reprogrammed through the later fixation of new genetic variants. Reprogramming phenomena in developmental pathways represent the basis of evolutionary change in phenotypes through the causation of new phenotypic variation. This variation is then sorted, with strong feedback to morphogenesis. One does not have to hypothesize radical reorganization of early morphogenetic systems as the cause of significant change of the phenotype. Small changes introduced early in pathways may have strong upwardly causing affects on the phenotype, most often as threshold phenomena. Because the relevant changes in early morphogenetic levels may be very small, because they may often be expressed in terms of change in quantitative parameters, and because of the highly integrated nature of developmental pathways in general, the chances are high that small early changes will in fact produce integrated rather than disruptive change in the phenotype. The resultant phenotype will not inevitably be a weird monster but rather a functioning organism. Not all changes in early development produce limbs in place of eyes, or even five-legged tetrapods (see van Valen, 1974). They produce a range of allowably different versions of the basic patterns, and thus there is a statistically useful likelihood that simple morphogenetic variations can produce (in addition to inevitable lethals) a set of alternative phenotypes that are immediately at least as functionally integrated as the original condition.

How then are morphogenetic pathways reprogrammed? Downward causation from late to earlier stages of morphogenesis seems unlikely in most cases. Morphogenesis seems almost always to be changed from the bottom up. Change in morphogenetic pathways can be introduced at any level and will have consequences for all subsequent levels. Morphogenesis may then become reprogrammed through the fixation of new genetic alleles and combinations in these later stages. The scale of potential change in phenotypes therefore depends in great part on the level of pattern control mechanism at which any change is initiated.

Heterochrony

Arguably one of the most simple types of reprogramming of morphogenesis is that occurring through change in the relative timing of different events. A change in timing immediately creates new conjunctions of existing elements of signal and response in a pathway or interacting pathways. "Loss" characters such as the absence of teeth in birds are readily explainable by change in developmental timing, in this case in epitheliomesenchymal interaction. We have noted previously that changes in complexity may result from a timing shift that sequesters part of a responding element so that not all of it receives the normal signal and is then left to acquire new properties at points further along the cascade. (This is a classic case of upward causation.)

Heterochronic mechanisms are particularly powerful when applied in late stages of pattern formation and affecting elements of relative size and shape of given organ systems. Much attention has been given to the interaction of the timing of size, shape, and sexual maturity as seen particularly by Gould (1977), whose elegant clock model neatly portrays the range of possible combinations of these three factors. Heterochrony, particularly in this last sense, is a particularly powerful and attractive model for evolutionary change because it requires no drastic reprogramming of the basic elements of the relevant morphogenetic pathways. The chances therefore are greatly increased that the new phenotype will be a fully integrated functional organism rather than the inevitable "lethal." A nice example here is the work of Hanken (1983a) on limb and skull modifications in "miniaturized" species of salamanders.

Alteration of the relative timing of size and shape change of something does not require a change in its basic "thingness" (to use an Aristotelean term). Heterochronic change producing the difference in cranial morphology between a chimpanzee and a human requires only the most subtle reprogramming of morphogenesis. The mandible is still the mandible; the first incisor is still the first incisor, and so on. However, the consequences may be considerable in those parameters where quantity rather than quality is crucial—the brain, for instance. A modest consequence in this shift is that the jaws may be too short for the full tooth row—leading, it is said, to the chronic human problem of the "wisdom teeth" having no room in which to erupt.

Heterochrony is thus a special case of reprogramming of development, one that always centers upon quantitative aspects of morphogenesis: size, rate, and timing in existing morphogenetic systems. It can, however, readily lead to qualitative shifts, especially if the timing change occurs early enough in a morphogenetic cascade to cause disruption of a subsequent stages.

Note on Macroevolution/Microevolution

This sort of analysis removes the old distinction between microevolutionary and macroevolutionary causes of evolutionary phenomena. In their place is a continuum of potential change, based on the hierarchical control in morphogenesis of different phenotypic character complexes.

Finally, the last argument that is usually summoned up to argue that rapid change of phenotypes through large-scale increments is impossible, is the old one: how does a hopeful monster find a mate and become propagated in a population? The answer to this is simple: the same way that any variant in a population does, and for the same reason.

CONSEQUENCES FOR SOME EVOLUTIONARY PROBLEMS

The fundamental difference between an approach to new kinds of evolutionary mechanisms through consideration of the role of morphogenetic processes and the traditional New Synthetic "microevolutionary" mechanisms is that the former is an internalist study whereas the latter sees evolution as driven by externalist causes. The internalist approach concentrates on the generation of new variation. This is not to say that internalist mechanisms drive all evolutionary change. Obviously, externalist mechanisms of selection and sorting are extremely important. But the externals must have something to work on. The aim of this book has been to concentrate on those internal morphogenetic properties that provide that raw material and to show that these generative processes play a major role in directing the course of evolution.

We have analyzed the control of the phenotype at different stages of morphogenesis and explored the evolutionary potential of morphogenetic systems in terms, principally, of the introduction of phenotypic variation among individual organisms. We may now turn to summarize these discussions briefly in terms of some long-standing evolutionary problems.

Rate and Scale of Morphological Change

One of the most difficult burdens for any discussion of the relationship between any aspect of development and evolution has been that of the "hopeful monster." In a model of evolution in which the driving force is an external sorting mechanism working on very fine-scale variations within populations, evolution must necessarily be slow and gradual. Nothing else seems possible; even multiple gene effects in populations will work slowly. Students of the role of development in evolution have therefore often been accused of conjuring up biologically meaningless "macromutations." Critic after critic (usually strong adherents of some other concept of evolutionary change) has made the claim that early development cannot be changed without lethality, that major shifts in the phenotype caused very quickly must be totally disruptive, or that significantly altered phenotypes would not find mates in the population in which they occurred. *Natura non agit per saltum.* In fact the argument almost seems to be that population genetics is more than sufficient for any imaginable rate of change; therefore, "macromutations" are unnecessary, and therefore they do not exist! However, comparison of the ontogenies of different organisms shows that early as well and late developmental pathways must in fact change and diverge. The nature of development is to produce highly integrated phenotypes. And any variant condition faces the same problems of propagation in populations. But it is principally a straw man because one does not need "hopeful monsters." Differential rates of evolutionary change of phenotypes, including rapid evolution, can be caused by quite normal processes and certainly without postulating leaps and bounds (of faith or morphology).

Perhaps the most important conclusion from the preceding discussions is that there is a hierarchy of control of different phenotypic character complexes that parallels a hierarchy of morphogenetic control of phenotypes. In the absence of a system of direct downward causation in morphogenesis, the internal contribution to evolutionary rates depends on the fact that different character complexes in the phenotypic hierarchy must have different inherent potentials for scale and rate of change. It is probably possible to disrupt any part of the morphogenetic system so as to produce small-scale changes in the relevant parts of the phenotype (although the course of change may be discontinuous). But it is only possible to produce a major evolutionary change in morphology (whether slowly or quickly) by means of changes in those characteristics that are controlled early in ontogeny, because that is where major pattern controls for phenotypic features operate. Late-stage morphogenetic controls affect only minor phenotypic character states.

In this case, rates of morphological and taxonomic change in evolution must be separable. High rates of taxonomic diversification are likely to come from a high rate of change in "minor" morphological and other characteristics (behavior, pheromones, etc.) that are based on changes caused in late developmental processes. These systems have the potential for a wide range of small-scale change within the bounds of developmental and functional constraints. But these changes cannot accumulate to produce major morphological change. That can only come from change in deeper morphogenetic levels. Here change is less likely to result in rapid taxonomic diversification. On the other hand, such "deep" changes are the basis of "progressive" evolutionary change: the evolution of new jaw mechanisms or limb patterns in vertebrates, new arrangements of the flowering parts in plants, or new segmental patternings in arthropods.

Variants caused at early morphogenetic levels will have the potential to affect major phenotypic systems through operation of fundamental pattern control mechanisms. A small initial change may produce a larger change after it has had a series of cascading effects along the morphogenetic pathway. A small initial change may also be reinforced rapidly through fixation of new alleles throughout the pathway. External processes of selection and sorting will be important, of course, but should not limit the rate of change in the lineage. This is an important difference between externalist and internalist approaches. The more phenotypes differ, the less they are bound by the same selective regime. Variants that are very similar will of course be sorted in the same environmental context. Variants that are significantly different have the chance to be tested on different terms; indeed, they make their own new terms.

Many authors have sought to find mechanisms for the production of (relatively rapid) innovation. Simpson (1953) talked of "quantum evolution," and Mayr (1963) hypothesized a "genetic revolution." Change initiated in early rather than late morphogenesis has the capacity to produce major changes in phenotypes over a short time period. It can do so because earlier changes will affect broader scales of control over the phenotype. The results will not inevitably be lethal because of the integrating and coordinating properties of morphogenesis. Changes will be rapid in some cases because of cascading effects within morphogenetic pathways and because of powerful feedback among different phenotypic levels (Chapter 8) and with the morphogenetic system. In such a hierarchical system, a given "amount" of (normally genetic) perturbation will have progressively different effects when introduced at different levels of the hierarchy. Such changes do not depend on a high rate of taxonomic diversification. Progressive evolution and taxonomic diversification are thus also uncoupled.

As for "hopeful monsters," all phenotypic variants are "hopeful monsters" thrown on the mercy of the environment. Once that is recognized, all that is left is an argument about scale. One should not confuse an argument about scale of production of variants ("all variants ought to be produced as extremely minor variations of existing forms") with an argument about what scale of change might be functionally acceptable. Again there is a simple problem of approach. If one starts from the premise that all organisms are optimally or maximally adapted, then one is forced to see any variant as having a lesser chance for survival. If, on the other hand, most organisms are less than optimally adapted and are merely coping, then any variant might have an equal or better chance. But the most important point is that any variant creates its own terms of reference and, precisely because it is different, it cannot by definition be judged in exactly the same terms as the sister or parent condition. Morphogenesis throws up a series of phenotypic variants: whether they survive or not will depend first on what they are.

Most discussions of evolutionary change in morphology concentrate on smoothly graded change. However, it is essential to face the fact that many evolutionary changes must have been binary. We have already discussed the case of the reptilian ankle joint configurations. Another is the evolution of the flatfish head with both eyes on the same side. It is only possible to think of this as a threshold phenomenon. The eye on one side could move within that side. The shift to the other side might involve an equally simple shift but one that has a discrete realm of new consequences.

Integration/Correlation of Morphological Change

A major problem in evolution is the explanation of apparently coordinated changes in highly integrated character complexes over significant lengths of time. On the one hand, one can trace many examples of this phenomenon, but on the other hand, traditional externalist mechanisms do not fully explain them. The answer obviously lies partially in the fact of developmental integration of these systems. The classic example is the continuity of association (Cuvier's "correlation of parts") of the set of bones that starts out as the jaw suspension of early fishes and ends up as the ossicles of the middle ear of mammals. At times during the history of these elements, a functional integration may be partially broken. In lower tetrapods the hyomandibular bone is already an ear ossicle (stapes), while the quadrate and articular bones are still firmly part of the jaw. Many arguments can be assembled to show that the jaw articulation in these forms is involved with sound transmission, and this may very well be true. But the fact is that evolution of jaw and hearing functions has perforce to act on elements of a complex of elements that is firmly linked in morphogenetic terms. The morphogenetic integrity is not broken.

The earlier the stage of morphogenesis at which a change is fixed, the more that change is likely to have a coordinated effect on other parts of the phenotype. However, there is an interesting element of coordination and integration of evolutionary change within the phenotype that comes from an integrating property of the context in which phenotypes are sorted. The environmental context does not merely act on isolated portions of the phenotype, but affects the whole. Thus, in the origin of land vertebrates from fishes the environment is selecting powerfully on the locomotor, hearing, feeding, (and other) systems all at the same time. This is where we can also find a powerful feedback situation where, through the coupling of different parts of the phenotype in morphogenesis and a common selective regime, change in one system will help reinforce particular sorts of change in another. The limiting factors will be the rate at which new genetic variation is introduced into the system and the extent to which morphogenetic pathways can be reprogrammed thereby. This feedback also extends to parts of the phenotype that are not closely linked developmentally—for example, the reproductive and feeding systems of fishes or the feeding and locomotor structures of horses, where there are shared externalist factors.

Origin and Diversification of Groups

The correlation of change among morphological character complexes is a special feature of both the origin and subsequent evolution of major groups. Perhaps the most obvious feature of major taxonomic groups of organisms is that we recognize them on the basis of major differences in patterns of morphology. The differences are rarely focused on a single morphological complex, and usually involve several character complexes. It has often seemed paradoxical that, judging from the fossil record, major groups seem to arise relatively suddenly with all their defining characteristics in place. These characteristics do not seem to be acquired gradually, in terms of either morphological scale or rate of change, as microevolutionary theory would predict. Similarly, the major new features seem to be acquired without the group having passed through major species-level diversifications, which would again be required in the gradual model.

The view of morphogenetic mechanisms discussed here offers an explanation of this paradox. The type of phenotypic variation (scale and rate) that normally characterizes microevolutionary change cannot accumulate directly to reprogram morphogenetic pathways. All stages of morphogenetic pathways are subject to change, and any change will have cascading effects along the pathway (upward causation). If major phenotypic characteristics are controlled at early rather than late morphogenetic stages, it is only through the initiation of change at the appropriate level that the sorts of innovations that characterize major groups can be produced. A single change or small number of changes at the appropriate level will produce the sort of effect that traditionally has been thought to require literally millions of generations of strong directed selection in microevolution and the rapid origin and extinction of hundreds of species. In this sense, macroevolution (the origin of major phenotypic shifts and/or high rate of morphological change) is mostly a matter of the hierarchical level at which evolutionary novelty is caused.

In terms of numbers of alleles fixed, the initial change in any major phenotypicshift may well be on the same order as that required for a small microevolutionary change. No macromutation is needed for such changes. Whereas a large phenotypic shift that depended on the accumulation of millions of small variants each expressed in different species would require a large number of genetic "events," a far smaller number is needed to achieve that effect by changing the control of early pattern formation mechanisms in morphogenesis. Furthermore, the differences in early morphogenetic mechanisms that distinguish major groups are small and subtle rather than large and overt. Then, as has already been stated several times, the highly integrated nature of developmental processes in general and morphogenesis in particular will ensure that those few small changes in the "right" place will not invariably produce a lethal effect but will occasionally produce a coordinated functioning phenotype. This phenotype is then tested in the environment in the usual way.

The second crucial feature of the model presented here is that the new phenotype will form the basis (theme) for a new set of variations. Progressive evolution therefore depends on the fixation of evolutionary changes at the right level in morphogenesis rather than the engendering of large taxonomic diversifications. Conversely (and perversely), most taxonomic diversifications add nothing to the general course of evolutionary change.

Trends

Evolutionary trends (including the once-popular notion of orthogenesis) have long presented a challenge for theory. One is bound to suspect that many examples of trends, particularly at lower taxonomic levels, are really taxonomists' artifacts, the product of the irresistible urge to find simple linear patterns rather than admit that nature often is a bit of a muddle.

Some well-known examples of trends include the directionality of limb modification or hyposodonty of the teeth in horse evolution, coiling of the shell in many invertebrates, and size increase in many clades. Their most obvious explanation would be causation through strong and historically consistent external directionality of selection. However, it is also obvious that many evolutionary trends also have a strong internal generative aspect. An obvious example would be any trend in shape that was driven by allometric size change. Some aspects of horse skull evolution are the direct consequence of increase in size. Size increase then is the real trend, and skull shape and proportion merely follows. But, as MacFadden (1986; cf. Radinsky, 1984) has shown, the early part of the history of horse evolution was accomplished in the absence of major size increase. Therefore, it is worth asking whether evolutionary trends may have some other deeper, internal cause.

While we can discount the possibility that trends are due to a bias in the initial introduction of genetic variations to a morphogenetic system, another cause of trends exists in terms of asymmetrical internal processes that sort positively for certain kinds of variation. We have seen that there is a potential for a powerful mechanism of positive feedback between external directional sorting of phenotypes and this internally operating sorting. One of the simplest ways in which such an open positive-feedback loop could operate is with respect to size increase, especially in a changing environmental context. Such open positive-feedback loops are the most likely explanation of the higher frequency verging on generality of evolutionary size increase in lineages (Cope's Law of Phyletic Size Increase). Such a feedback system would also explain many cases of trends in the correlated progression of change in phenotypic characters.

In addition, trends may be the result of developmental constraints. If the morphogenetic system is constrained in a way that produces an asymmetry between the introduction of genetic and phenotypic variation, and if the mechanism underlying the constraint is itself unmodified during the course of evolutionary change in particular characters, then it will continue to operate in the same direction over a long time period. In the vertebrate limb, for example, there seem to be fundamental constraints involved in the apportioning of mesenchyme within the limb bud. A whole range of modifications to the control of morphogenetic events will give the same phenotypic results: loss of the lateral digits before the axial ones, reduction of the zeugopodium before the stylopodium. Many different genetic variations, introduced at different levels in the morphogenetic cascade, could combine under this "integrative" influence to cause phenotypic changes in the form of a "trend."

Parallelism and Convergence

Closely allied with "trends" are parallelism and convergence. Here again a developmental mechanism will explain the phenomenon better than a mechanism that depends on selection alone. Parallelism and convergence will occur when the same fundamental component of a developmental system is disrupted. Thus, for example, both the perissodactyl and artiodactyl ungulates show a "tendency" for lateral digit reduction. They show a different pattern of limb modification (one digit versus two), but the basic nature of the modification is much the same, including the formation of the hooves. Even more dramatically, the South American Litopterna show a striking convergence on the horse condition, down even to the evolution of single-toed species. It would be naive to propose that all these lineages have somehow acquired huge suites of the identical genes to accomplish this. It is not necessary to propose viral-mediated DNA transfer from lineage to lineage (even though this may be reasonable in some instances; cf. Jeppsson, 1986). Instead (to repeat), the evidence from morphogenetic pattern control of limb mutants and other systems shows that the same phenotypic effect can be caused through disruption of a particular element of the developmental hierarchy by several different causes. What parallel and convergent systems have in common is the level of mechanism in the developmental hierarchy that is disrupted, not the particular "mutation" that actually does the job. According to the model suggested here, parallelism in ungulate digit reduction would be the consequence of any of a potentially large number of genetically or epigenetically mediated perturbations of the pattern control of mesenchyme distribution in the early limb bud. Parallelism and convergence are, in fact, inevitable in a hierarchically arranged set of developmental mechanisms. They represent, as it were, the inverse of developmental constraints. They are not predictable from a study of individual genes and gene expression and, except at a phenotypically trivial level, neither predicted nor explained in terms of population genetics.

POSTSCRIPT—RESEARCH PROSPECTUS

Much of what has been analyzed here and much of the discussion has been conjectural. I hope that it has been more than enough to show the potential for the internal nature of morphogenetic systems to add a layer to evolutionary processes. But much work remains to be done.

First and foremost, there is still a great deal that we do not know about the details of morphogenetic mechanisms. This gap has been filling fast in the last 5 years. The data will inevitably come in piecemeal—some molecular aspects here, some organ-level work there. An important task will be to use all the information we can get to assemble more complete accounts of morphogenetic cascades, rather than simply describe isolated morphogenetic events. One area where we can hope for good results is in the case of morphogenesis of the tetrapod limb, from molecular mechanisms in the early limb bud to remodeling of ossifications in adults. An especially important aspect of this type of integration will be to study associated cascades and the buildup of complex morphological character complexes—for example, the interrelationships in morphogenesis of muscles, ligaments, vascularization, and innervation of the limb in association with the pattern of skeletal elements. To do this we need to find both the driving generalities of pattern control and the particularities of given subsystems.

We need to know much more about how morphogenetic cascades change. This can be accomplished in part by careful analyses of the differences in pathways between organisms of known phylogenetic relationship. It must also be approached experimentally. Experimental work is also especially needed to study how and why morphogenetic pathways do not change, or resist change. That is to say we need to study the morphogenetic basis of the phenomenon of canalization and the perennial question of constraint.

Developmental constraint is one of the most familiar concepts in this whole subject. But while we have a good grasp of what it is, in the sense of what it produces, we have a very limited view of what causes constraints in particular systems. For example, a constant allometric relationship in development is an example of a constrained feature. We do not know the mechanisms that cause the constancy of allometric relationships within clades, or why some allometries are apparently more easily changed than others. What is entailed in causing an allometric coefficient to change? Is there some general principle, or merely a set of unique contingent events for each taxon?

Finally, one of the most exciting possible areas for future research is in the ecological control of morphogenesis and indeed all development. Here we can bring together important externalist and internalist approaches to look at questions such as the causal links between functional and developmental integration.

Alfred Sherwood Romer once wrote (1966) that "in vertebrate paleontology, increasing knowledge leads to triumphant loss of clarity." This is probably true of all fields at some time or another. However, there are also phases in the history of a particular science when things start to come together, when long sought-for general principles begin to link disciplines, instead of different techniques and masses of underanalyzed data tending to drive them apart. Twenty years ago evolutionary biology was becoming something of a dull subject in which all the questions seemed to have been asked and most of them had been answered. But a lot of approaches had been excluded. There then followed a period of attack from several sides; new approaches brought conflict, and old data were now confusingly not explained. All the current signs point to a new period of synthesis where diverse biological disciplines may once again be brought together and agreement may be possible on the next questions to be asked. To be sure, what had become simplified has become complex and controversial again, but the result is as exciting as the possibilities seem endless.

References

Alberch, P. (1983). Morphological variation in the neotropical salamander genus *Bolitoglossa*. *Evolution* 37:906–919.

Alberch, P., and Gale, E.A. (1985). A developmental analysis of an evolutionary trend: Digital reduction in amphibians. *Evolution* 39:8–23.

Ambros, V., and Horvitz, H.R. (1984). Heterochromic mutants of the nematode *Caenorhabditis elegans*. *Science* 226:409–416.

Andersen, C.B., and Meier, S. (1981). The influence of the metameric pattern in the mesoderm on the migration of cranial neural crest cells in the chick embryo. *Dev. Biol.* 85:385–402.

Archer, C.W., Hornbruch, A., and Wolpert, L. (1983). Growth and morphogenesis of the fibula in the chick embryo. *J. Embryol. Exp. Morphol.* 75:101–106.

Archer, C.W., Rooney, P., and Cottrill, C.P. (1985). Cartilage morphogenesis in vitro. *J. Embryol. Exp. Morphol.* 90:33–44.

Arnold, S.M. (1968). The role of the egg cortex in cephalopod development. *Dev. Biol.* 18:180–197.

Arnolds, W.J.A., van der Biggelow, J.A.M., and Verdonk, N.H. (1983). Spatial aspects of cell interactions involved in the determination of dorso-ventral polarity in equally cleaving gastropods and regulative abilities of their embryos as studied by micromere deletions in *Lymnaea* and *Patella*. *Wm. Roux's Arch. Dev. Biol.* 192:78–86.

Bachmann, K. (1983). Evolutionary genetics and the genetic control of morphogenesis in flowering plants. *Evol. Biol.* 16:157–208.

Baron, R., Tross, R., and Vignery, A. (1984). Evidence of sequential remodeling in rat trabecular bone: Morphology, dynamic histomorphometry, and changes during skeletal maturation. *Anat. Rec.* 208:137–145.

Bateson, G. (1979). *Mind and Nature.* Dutton, New York.

Beckner, M. (1974). Reduction, hierarchies and organisms. In F.J. Ayala and T. Dobzhansky (eds.), *Studies in the Philosophy of Biology*. Univ. of California Press, Berkeley.

de Beer, G.R. (1958). *Embryos and Ancestors,* 3rd ed. Oxford Univ. Press (Clarendon), London/New York.

Black, S.D., and Gerhart, J.C. (1985). Experimental control of the site of embryonic axis formation in *Xenopus laevis* eggs centrifuged before first cleavage. *Dev. Biol.* 108:310–324.

Bolton, V.N., Oades, P.J., and Johnson, M.H. (1984). The relationship between cleavage, DNA replication, and gene expression in the mouse 2-cell embryo. *J. Embryol. Exp. Morphol.* 79:139–163.

Bonner, J.T. (ed.) (1982). *Evolution and Development.* Springer-Verlag, Berlin, Heidelberg, New York.

Boucaut, J., and Darribiere, T. (1983). Presence of fibronectin during early embryogenesis in amphibian *Pleurodeles waltlii*. *Cell Differ.* 12:77–83.

Boucaut, J.C., Darribiere, T., Boulekbecka, H., and Thiery, J.P. (1986). Prevention of gastrulation but not neurulation by antibodies to fibronectin in amphibian embryos. *Nature* 307:364–366.

Brauer, P.R., Bolender, D.L., and Markwald, R.R. (1983). Localization of 3H-fucosylated substances in the pathway of migrating cephalic neural crest cells. In P.W. Coates, R.R.

Markwald, and A.D. Kenny (eds.), *Developing and Regenerating Vertebrate Nervous Systems*. Alan R. Liss, New York.

Briggs, R., and King, T.S. (1952). Transplantation of living nuclei from blastula cells into enucleated frog's eggs. *Proc. Natl. Acad. Sci. USA* 38:455–463.

Bronner-Fraser, M. (1982). Distribution of latex beads and retinal pigment cells along the ventral neural crest pathway. *Dev. Biol.* 91:50–63.

Bronner-Fraser, M. (1986). Analysis of the early stages of trunk neural crest migration in avian embryos using monoclonal antibody HNK-1. *Dev. Biol.* 115:44–55.

Campbell, J.H. (1982). Autonomy in evolution. In R. Milkman (ed.), *Perspectives on Evolution*. Sinauer, Sunderland, Mass.

Campbell, T. (1974). Downward causation in hierarchically organized biological systems. In J.J. Ayala, and T. Dobzhansky (eds.), *Studies in the Philosophy of Biology*. Univ. of California Press, Berkeley.

Capco, D.G., and McGaughey, R.W. (1986). Cytoskeletal reorganization during early mammalian development; Analysis using embedment-free sections. *Dev. Biol.* 115:446–458.

Chevalier, A. (1978). Étude de la migration des cellules somatiques dan le mesoderme somatopleural de l'ebauche de l'aile. *Wm. Roux' Arch. Dev. Biol.* 184:57–73.

Christ, B., Jacob, H.J. and Jacob, M. (1977). Experimental analysis of the origin of the wing musculature in avian embryos. *Anat. Embryol.* 150:171–186.

Clarke, G.M., and McKenzie, J.A. (1987). Developmental stability of insecticide-resistant phenotypes in blowfly: A result of canalising natural selection. *Nature* 325:345–346.

Cleine, J.H., and Slack, J.M.W. (1985). Normal fates and states of specification of different regions in the axolotl gastrula. *J. Embryol. Exp. Morphol.* 86:247–269.

Cohen, J. (1979). Maternal constraints on development. In D.R. Newth and M. Balls (eds.), *Maternal Effects in Development*. Cambridge Univ. Press, London/New York.

Cooke, J., and Webber, J.A. (1983). Vertebrate embryos: Diversity in developmental strategies. *Nature* 306:423–424.

Cooke, J., and Webber, J.A. (1985). Dynamics of the control of body pattern in the development of *Xenopus laevis*. *J. Embryol. Exp. Morphol.* 88:85–133.

Curtis, A.S.G.(1960). Cortical grafting in *Xenopus laevis*. *J. Embryol. Exp. Morphol.* 10:410–422.

Curtis, A.S.G. (1962). Morphogenetic interactions before gastrulation in the amphibian, *Xenopus laevis*: The cortical field. *J. Embryol. Exp. Morphol.* 10:420–422.

Dale, L., Smith, J.C., and Slack, J.M.W. (1985). Mesoderm induction in *Xenopus laevis*: A quantitative study using a cell lineage label and tissue-specific antibodies. *J. Embryol. Exp. Morphol.* 89:289–312.

Darwin, C.R. (1859). *On the Origin of Species*. Murray, London.

Duband, J.L., and Thiery, J.P. (1982). Appearance and distribution of fibronectin during chick embryo gastrulation and neurulation. *Dev. Biol.* 94:337–350.

du Brul, E.L., and Laskin, D.M. (1961). Preadaptive potentialities in the mammalian skull; an experiment in growth and form. *Am. J. Anat.* 109:117–132.

Ede, D.A. (1976). Cell interaction in vertebrate limb development. In G. Poste and G.L. Nicholson (eds.), *The Cell Surface in Animal Embryogenesis and Development*. Elsevier, Amsterdam/New York.

Ekaratne, S.V.K., and Crisp, D.J. (1983). A genetic analysis of torsion in gastropod shells, with particular reference to turbinate forms. *J. Mar. Biol. Assoc. UK*, 63:777–797.

Eldredge, N. (1983). Phenomenological levels and evolutionary rates. *Syst. Zool.* 31:338–347.

Eldredge, N., and Stanley, S. (1984). *Living Fossils*. Springer-Verlag, New York/Berlin.

Eldredge, N., and Salthe, S.N. (1985). Hierarchies and evolution. *Oxford Surv. Evol. Biol.* 1:184–208.

Elinson, R.P. (1980). The amphibian egg cortex in fertilisation and early development. In S. Subtelny and N.K. Wessels (eds.), *Cell Surface: Mediator of Developmental Processes*. *Symp. Soc. Dev. Biol.* 38:217–234.

Erickson, C.A. (1985). Control of neural crest cell dispersion in the trunk of the avian embryo. *Dev. Biol.* 111:138–157.

Fallon, J.F., and Cameron, J. (1977). Interdigital cell death during limb development of the turtle and lizard with an interpretation of evolutionary significance. *J. Embryol. Exp. Morphol.* 40:285–289.

Fleming, T.P., Cannon, P.M., and Pickering, S.J. (1986). The cytoskeleton, endocytosis and cell polarity in the mouse preimplantation embryo. *Dev. Biol.* 113:406–419.

Franke, W.W., Grund, C., Jackson, B.W., and Illmensee, K. (1983). Formation of cytoskeletal elements during mouse embryogenesis. *Differentiation* 25:121–141.

French, V., Bryant, P.J., and Bryant, S.V. (1976). Pattern regulation in epimorphic fields. *Science* 193:969–981.

Galau, G.A., Klein, W.H., Davis, M.M., Wold, B.J., Britten, R.J., and Davidson, E.H. (1976). Structural genes sets active in embryos and adult tissues of the sea urchin. *Cell* 7:487–505.

Gautier, J., and Beetschen, J. (1985). A three-step scheme for gray crescent formation in the rotated axolotl oocyte. *Dev. Biol.* 110:192–199.

Geist, V. (1978). *Life Strategies, Human Evolution, Environmental Design: Toward a Biological Theory of Health.* Springer-Verlag, New York/Berlin.

Gerhart, J.C., Black, S., and Scharf, S. (1983). Cellular and pancellular organization of the amphibian embryo. *Mod. Cell Biol.* 2:483–507.

Gerhart, J.C., Ubbels, G., Black, S., and Kirschner, M. (1981). A reinvestigation of the role of the gray crescent in axis formation in *Xenopus laevis. Nature* 292:511–516.

Ghiselin, M.T. (1974). A radical solution to the species problem. *Syst. Zool.* 25:536–544.

Ghiselin, M.T. (1987). Hierarchies and their components. *Paleobiology* 13:108–111.

Gimlich, R.L., and Gerhart, J.C. (1984). Early cellular interactions promote embryonic axis formation in *Xenopus laevis. Dev. Biol.* 104:117–130.

Goldschmidt, R.B. (1940). *The Material Basis of Evolution.* Yale Univ. Press, New Haven, Conn.

Goodday, D., and Thorogood, P.V. (1985). Contact behaviour exhibited by migrating neural crest cells in confrontation culture with somitic cells. *Cell. Tissue Res.* 241:165–169.

Goodrich, E.S. (1930). *The Structure and Development of Vertebrates.* Macmillan & Co., London.

Goodwin, B.C. (1982). Development and evolution. *J. Theor. Biol.* 85:757–770.

Goodwin, B.C. (1984). Changing from an evolutionary to a generative paradigm in biology. In J.W. Pollard (ed.), *Evolutionary Theory.* Wiley, New York.

Goodwin, B.C., and Cohen, M.H. (1969). A phase-shift model for the spatial organization of developing systems. *J. Theor. Biol.* 25:49–107.

Goodwin, B.C., and Trainor, J.E.H. (1983). A field description of the cleavage process in embryogenesis. In B.C. Goodwin, N. Holder, and C.C. Wylie (eds.), *Development and Evolution.* Cambridge Univ. Press, Cambridge.

Goodwin, B.C., Holder, N., and Wylie, C.C. (eds.) (1983). *Development and Evolution.* Cambridge Univ. Press, Cambridge.

Gordon, J.B. (1977). Egg cytoplasm and gene control in development. *Proc. R. Soc. Lond.* 198:211–247.

Gould, S.J. (1973). Positive allometry of antlers in the "Irish Elk," *Megaloceros giganteus. Nature* 244:375–376.

Gould, S.J. (1977). *Ontogeny and Phylogeny.* Harvard Univ. Press, Cambridge, Mass.

Gould, S.J. (1984). Morphological channeling by structural constraint. *Paleobiology* 10:172–194.

Gould, S.J., and Eldredge, N. (1977). Punctuated equilibria: The tempo and mode of evolution reconsidered. *Paleobiology* 3:115–151.

Gould, S.J., and Lewontin, R.C. (1979). The spandrels of San Marcos and the Panglossian paradigm: A critique of the adaptationist program. *Proc. Roy. Soc. London* B 205:581–598.

Gould, S.J., and Vrba, E.S. (1982). Exaptation—a missing term in the science of form. *Paleobiology* 8:4–15.

Graham, C.F., and Waering, P.F. (eds.) (1984). *Developmental Control in Animals and Plants,* (2nd edition). Blackwell, Oxford.

Grene, M. (1983). *Dimensions of Darwinism.* Cambridge Univ. Press, London/New York.

Grene, M. (1987). Hierarchies in biology. *Am. Sci.* 75:506–510.

Gruneberg, H. (1963). *The Pathology of Development*. Blackwell, Oxford.

Guthrie, D.M., and Banks (1970). Observations on the function and physiological properties of a fast paramyosin muscle—the notochord of amphioxus (*Branchiostoma lanceolatum*). *J. Exp. Biol.* 52:125–138.

Haeckel, E. (1866). *Naturliche Schöpfungsgeschichte*. Reimer, Berlin.

Hall, B.J. (1984). Developmental mechanisms underlying the formation of atavisms. *Biol. Rev.* 59:89–124.

Hampe, A. (1959). Contribution a l'etude du development et de la regulation des difficiences et des excedents dans la platte de l'embryon de Poulet. *Arch. Anat. Microsc. Morphol. Exp.* 48:345–478.

Hampe, A. (1960). Le competition entre les elements osseux du zeugopode de Poulet. *J. Embryol. Exp. Morphol.* 8:241–245.

Hanken, J. (1983a). Genetic variation in a dwarfed lineage, the Mexican salamander genus *Thorius* (Amphibia: Plethodontidae): Taxonomic, ecologic and evolutionary implications. *Copeia* 4:1051–1073.

Hanken, J. (1983b). High incidence of limb skeletal variants in a peripheral population of the red-backed salamander *Plethodon cinereus* (Amphibia: Plethodontidae), from Nova Scotia. *Can. J. Zool.* 61:1925–1931.

Hanken, J. (1985). Morphological novelty in the limb skeleton accompanies miniaturization in salamanders. *Science* 229:871–873.

Hay, E.D. (1983). Cell and extracellular matrix: Their organization and mutual dependence. *Mod. Cell Biol.* 2:509–548.

Helfer, S.R., and Helfer, E.S. (1983). Computer simulation of organogenesis: An approach to the analysis of shape changes in epithelial organs. *Dev. Biol.* 97:444–453.

Herschel, J.F.W. (1830). *A Preliminary Discourse on the Study of Natural Philosophy*. Longmans, London.

Hinchliffe, J.R., and Thorogood, P.V. (1974). Genetic inhibition of mesenchymal cell death and the development of form and skeletal pattern in the limbs of talpid 3 mutant chick embryos. *J. Embryol. Exp. Morphol.* 31:747–760.

Hinchliffe, J.R., and Johnson, D.R. (1980). *The Development of the Vertebrate Limb*. Oxford Univ. Press (Clarendon), London/New York.

Hinchliffe, J.R. and Griffiths, P.J. (1983). The prechondrogenic patterns in tetrapod limb development and their phylogenetic significance. In B.C. Goodwin, N. Holder, and C.C. Wylie (eds.), *Development and Evolution*. Cambridge Univ. Press, London/New York.

Ho, M-W., and Saunders, P.T. (1979). Beyond neo-Darwinism: An epigenetic approach to evolution. *J. Theor. Biol.* 78:573–591.

Hodge, M.J.S. (1977). The structure and strategy of Darwin's "Long Argument." *Br. J. Hist. Sci.* 10:237–246.

Holder, N. (1983a). Developmental constraints and the evolution of vertebrate digit patterns. *J. Theor. Biol.* 104:451–471.

Holder, N. (1983b). The vertebrate limb: Patterns and constraints in development and evolution. In B.C. Goodwin, N. Holder, and C.C. Wylie (eds.). *Development and Evolution*. Cambridge Univ. Press, London/New York.

Horder, T.J. (1983). Embryological bases of evolution. In B.C. Goodwin, N. Holder, and C.C. Wylie (eds.), *Development and Evolution*. Cambridge Univ. Press, London/New York.

Horstadius, S. (1973). *Experimental Embryology of Echinoderms*. Oxford Univ. Press, New York/London.

Hull, D.L. (1980). Individuality and selection. *Annu. Rev. Ecol. Syst.* 11:311–332.

Hurle, J.M., and Ganon, Y. (1986). Interdigital tissue chondrogenesis induced by surgical removal of the ectoderm in the embryonic chick leg bud. *J. Embryol. Exp. Morphol.* 94:231–244.

Hutchinson, G.E. (1981). Random adaptation and imitation in human evolution. *American Scientist* 69:161–165.

Huxley, J.S. (1932). *Problems of Relative Growth*. Methuen, London.

Huxley, J.S., and de Beer, G.R. (1934). *The Elements of Experimental Embryology*. Cambridge Univ. Press, London/New York.

Jacob, H.J., and Christ, B. (1980). On the formation of muscular patterns in the chick limb. In J. Merker (ed.), *Teratology of the Limb*. De Gruyter, Berlin.

Jacobson, A.G., Oster, G.F., Odell, G.M., and Cheny, L.Y. (1986). Neurulation and the cortical tractor model for epithelial folding. *J. Embryol. Exp. Morphol.* 96:19–49.

Jacobson, M. (1982). Origin of the nervous system in amphibians. In N.C. Spitzer (ed.), *Neuronal Development*. Plenum, New York.

Jacobson, M. (1984). Cell lineage analysis of neural induction: Origin of cells forming the induced nervous system. *Dev. Biol.* 102:122–129.

Jeffery, W.R. (1985). Identification of proteins and mRNAs in isolated yellow crescents of ascidian eggs. *J. Embryol. Exp. Morphol.* 89:275–287.

Jeffery, W.R., and Meier, S. (1983). A yellow crescent cytoskeletal domain in ascidian eggs and its role in early development. *Dev. Biol.* 96:125–143.

Jepsen, G.L., Mayr, E., and Simpson, G.G. (1949). *Genetics, Paleontology and Evolution*. Princeton Univ. Press, Princeton, N.J.

Jeppsson, L. (1986). A possible mechanism of convergent evolution. *Paleobiology* 12:80–88.

Johnson, M.H., Ziomek, C.A., Reeve, W.J.D., Pratt, H.P.M., Goodall, H., and Handyside, A.H. (1984). The mosaic organisation of the preimplantation mouse embryo. In J. Van Blerkom, and P.M. Motta, (eds.), *Ultrastructure of Reproduction*. Martinus Nijhoff, Boston, The Hague, Dordrecht, Lancaster.

Kageura, H., and Yamana, K. (1983). Pattern regulation in isolated halves and blastomeres of early *Xenopus laevis*. *J. Embryol. Exp. Morphol.* 74:221–234.

Kao, K.R., and Elinson, R.P. (1985). Alteration of the anterior-posterior embryonic axis: The pattern of gastrulation in macrocephalic frog embryos. *Dev. Biol.* 107:239–251.

Kauffman, S.A. (1974). The large-scale structure and dynamics of gene control circuits: an ensemble approach. *J. Theor. Biol.* 44:167–190.

Kauffman, S.A. (1983). Developmental constraints: Internal factors in evolution. In B.C. Goodwin, N. Holder, and C.C. Wylie (eds.), *Development and Evolution*. Cambridge Univ. Press, London/New York.

Kauffman, S.A., Shmyko, R.M., and Trabert, K. (1978). Control of sequential compartment formation. *Science* 199:289–270.

Kay, E.D. (1986). The phenotypic interdependence of the musculoskeletal characters of the mandibular arch in mice. *J. Embryol. Exp. Morphol.* 98:123–136.

Keller, R.E. (1978). Time-lapse cinematographic analysis of superficial cell behavior during and prior to gastrulation in *Xenopus laevis*. *J. Morphol.* 157:223–248.

Keller, R.E. (1980). The cellular basis of epiboly: An SEM study of deep cell rearrangements during gastrulation of *Xenopus laevis*. *J. Embryol. Exp. Morphol.* 60:201–234.

Keller, R.E. (1981). An experimental analysis of the role of bottle cells and the deep marginal zone in the gastrulation of *Xenopus laevis*. *J. Exp. Zool.* 216:81–101.

Kemp, R.B., and Hinchliffe, J.R. (eds.) (1984). *Matrices and Cell Differentiation*. Alan R. Liss, New York.

Kemp, T.S. (1982). *Mammal-Like Reptiles and the Origin of Mammals*. Academic Press, New York/London.

Kemp, T.S. (1985). Synapsid reptiles and the origin of higher taxa. *Spec. Pap. Paleont.* 33:175–184.

Kimura, M. (1983). *The Neutral Theory of Molecular Evolution*. Cambridge Univ. Press, London/New York.

King, T.J., and Briggs, R. (1956). Serial transplantation of embryonic nuclei. *Cold Spring Harbor Symp. Quant. Biol.* 21:271–289.

Kirschner, M.W., and Gerhart, J.C. (1981). Spatial and temporal changes in the amphibian egg. *Bioscience* 31:381–388.

Kollar, E.J., and Fisher, C. (1980). Tooth induction in chick epithelium. *Science* 207:993–995.

Korn, L.J. (1982). Transcription of *Xenopus* 5S ribosomal RNA genes. *Nature* 295:101–105.

Kosher, R.A., Gay, S.W., Kamanitz, J.R., Kulyk, W.M., Rodger, B.J., Sai, S., Tanake, T., and Tanzer, M.L. (1986). Cartilage proteoglycan core protein gene expression during limb cartilage differentiation. *Dev. Biol.* 118:112–117.

Lauder, G.V. (1981). Form and function: Structural analysis in evolutionary biology. *Paleobiology* 7:430–442.

Lauder, G.V. (1982). Historical biology and the problem of design. *J. Theor. Biol.* 97:57–67.

Levinton, J.S. (1985). Developmental constraints and evolutionary saltation: A discussion and critique. In J.P. Gustafson and F.J. Ayala (eds.), *Genetics, Development and Evolution.* Univ. of Missouri Press, Columbia.

Lewis, E.B. (1978). A gene complex controlling segmentation in *Drosophila. Nature* 276:565–570.

Lewontin, R.C. (1970). The units of selection. *Ann. Rev. Ecol. Syst.* 1:1–16.

Liem, K.F. (1973). Evolutionary strategies and morphological innovations: Cichlid pharyngeal jaws. *Syst. Zool.* 22:425–441.

Loer, C.M., Steeves, J.D., and Goodman, C.S. (1983). Neuronal cell death in grasshopper embryos: Variable patterns in different species, clutches, and clones. *J. Embryol. Exp. Morphol.* 78:169–182.

Macbeth, N. (1971). *Darwin Retried.* Gambit, Boston.

MacFadden, B.J. (1986). Fossil horses from *"Eohippus"* to *Equus*: scaling, Cope's Law, and the evolution of body size. *Paleobiology* 12:355–369.

Malacinski, G.M., Chung, H.M., and Ashima, M. (1980). The association of primary embryonic organiser activity with the future dorsal side of amphibian eggs and early embryos. *Dev. Biol.* 77:449–462.

Mallatt, J. (1984). Feeding ecology of the earliest vertebrates. *Zool. J. Linn. Soc. Lond.* 82:261–272.

Marthy, W.J. (1975). Organogenesis in Cephalopoda: Further evidence of blastodisc-bound developmental information. *J. Embryol. Exp. Morphol.* 33:75–88.

Martin, P., and Lewis, J. (1986). Normal development of the skeleton in chick limb buds devoid of dorsal ectoderm. *Dev. Biol.* 118:233–246.

Mayr, E. (1963). *Animal Species and Evolution.* Harvard Univ. Press, Cambridge, Mass.

Meckel, J.F. (1821). *System der vergleichenden Anatomie.* Rengerschend Buchhandlung, Halle.

Medawar, P.B. (1954). The significance of inductive relationships in the development of vertebrates. *J. Embryol. Exp. Morphol.* 2:172–174.

Meier, S., and Drake, C. (1984). SEM localization of cell-surface-associated fibronectin in the cranium of chick embryos utilizing immunolatex microspheres. *J. Embryol. Exp. Morphol.* 80:175–195.

Meinhardt, H. (1982). *Models of Biological Pattern Formation.* Academic Press, London/New York.

Meuler, D.C., and Malacinski, G.M. (1985). An analysis of protein synthesis patterns during early embryogenesis of the urodele. *J. Embryol. Exp. Morphol.* 89:71–92.

Morrill, J.B., Blair, C.A., and Larsen, W.J. (1973). Regulative development in the pulmonate gastropod *Lymnea palustris*, as determined by deletion experiments. *J. Embryol. Exp. Morphol.* 183:47–56.

Muller, G. (1985). Experimentelle Untersuchungen zur Theorie des epigenetischen Systems. In S.A. Ott, G.P. Wagner, and F.M. Wuketis (eds.), *Evolution, Ordnung und Erkenntnis.* Parey, Berlin.

Muller, G. (1986). Effects of skeletal change on muscle pattern formation. *Bibliotheca Anatomica* 29:91–108.

Muller, G., and Wagner, G. (1987). Comparative anatomy of experimentally produced atavisms in the chick hindlimb: An evolutionary epigenetic approach. In press.

Murray, J.D. (1981). A pre-pattern formation mechanism for animal coat markings. *J. Theor. Biol.* 88:161–199.

Nagel, E. (1961). *The Structure of Science.* Harcourt, Brace & World, New York.

Nakamura, O. (1978). The epigenetic formation of the organizer. In O. Nakamura, and S. Toivonen, (eds.), *The Organizer.* Elsevier–North Holland, Amsterdam/New York.

Nakamura, O., and Matsuzawa, T. (1967). Differentiation capacity of the dorsal marginal zone in the morula and blastula of *Triturus pyrrhogaster. Embryologica* 9:223–237.

Nakamura, O., and Takasaki, H. (1970). Further studies on the differentiation capacities of the dorsal marginal zone in the morula of *Triturus pyrrhogaster. Proc. Jpn. Acad.* 46:546–551.

Nakamura, O., and Toivonen, S. (1978). *The Organiser.* Elsevier–North Holland, Amsterdam/New York.

Nakatsuji, N., and Johnson, K.E. (1983). Conditioning of a culture substratum by the ectodermal layer promotes attachment and oriented locomotion by amphibian gastrula mesodermal cells. *J. Cell. Sci.* 59:43–60.

Nakatsuji, N., Smolira, M.A., and Wylie, C.C. (1985). Fibronectin visualized by scanning electron microscopy immunocytochemistry on the substratum for cell migration in *Xenopus laevis* gastrulae. *Dev. Biol.* 107:264–268.

Neff, A.W., Wakahara, M., Jurand, A., and Malacinski, G.M. (1984). Experimental analyses of cytoplasmic rearrangements which follow fertilization and accompany symmetrization of inverted *Xenopus* eggs. *J. Embryol. Exp. Morphol.* 80:197–224.

Neff, A.W., Malacinski, G.M., and Chung, H. (1985). Microgravity simulation as a probe for understanding early *Xenopus* pattern specification. *J. Embryol. Exp. Morphol.* 89:259–274.

Newton-Smith, W.H. (1982). *The Rationality of Science.* Routledge & Kegan Paul, London.

Nieuwkoop, P.D. (1973). The "organisation center" of the amphibian embryo: its origin, spatial organisation and morphogenetic action. *Adv. Morphogenetics* 10:1–39.

Noden, D.M. (1983). The role of the neural crest in patterning avian cranial, skeletal, connective and muscle tissues. *Dev. Biol.* 96:144–165.

Noden, D.M. (1986). Inherited homeotic midfacial malformations in Burmese cats. *Craniofacial Genet. and Dev.Biol.* (Suppl.) 2:249–266.

Northcutt, R.G., and Gans, C. (1983). The origin of neural crest and epidermal placodes: A reinterpretation of vertebrate origins. *Q. Rev. Biol.* 58:1–28.

Odell, G., Oster, G.F., and Alberch, P. (1980). Mechanisms, morphogenesis and evolution. In G. Oster (ed.), *Lectures on Mathematics in the Life Sciences.* American Mathematics Society, Providence, R.I.

Odell, G., Oster, G., Alberch, P., and Burnside, B. (1981). The mechanical basis of morphogenesis I: A model for epithelial tissue folding. *Dev. Biol.* 85:446–462.

O'Grady, R.T. (1984). Evolutionary theory and teleology. *J. Theor. Biol.* 107:563–578.

Oster, G., and Alberch, P. (1982). Evolution and bifurcation of developmental programs. *Evolution* 36:444–459.

Oster, G.F., Murray, J.D., and Harris, A.K. (1983). Mechanical aspects of mesenchymal morphogenesis. *J. Embryol. Exp. Morphol.* 78:83–125.

Oster, G.F., Murray, J.D., and Maini, P.K. (1985). A model for chondrogenic condensations in the developing limb: The role of extracellular matrix and cell tractions. *J. Embryol. Exp. Morphol.* 89:93–112.

Owens, E.M., and Solursh, M. (1983). Accelerated maturation of limb mesenchyme by the Brachypod H mouse mutation. *Differentiation* 24:145–148.

Pasteels, J. (1948). Les bases de la morphogenese chez les vertebres anamniotes au function de la structure de l'oeuf. *Folia Biotheretica* 3:83–108.

Paterson, H.E.H. (1978). More evidence against speciation by reinforcement. *S. Afr. J. Sci.* 74:369–371.

Patou, M.P. (1977). Dorso-ventral axis determination of chick limb bud development. In D.A. Ede, J.R. Hinchliffe, and M. Balls (eds.), *Vertebrate Limb and Somite Morphogenesis.* Cambridge Univ. Press, London/New York.

Patterson, C. (1983). How does phylogeny differ from ontogeny? In B.C. Goodwin, N. Holder, and C.C. Wylie (eds.), *Development and Evolution.* Cambridge Univ. Press, London/New York.

Platt, J.B. (1893). Ectodermic origin of the cartilages of the head. *Anat. Anz.* 8:506–509.

Provasoli, L., and Pintner, I.J. (1980). Bacteria induced polymorphism in an axenic laboratory strain of *Ulva lactuca* (Chlorophysceae). *J. Phycol.* 16:196–201.

Rachootin, S.P., and Thomson, K.S. (1981). Epigenetics, palaeontology and evolution. In G.G.E. Scudder, and C.L. Reveal (eds.), *Evolution Today*. Hunt Institute, Pittsburgh, Pa.

Radinsky, L. (1984). Ontogeny and phylogeny in horse skull evolution. *Evolution* 38:1–15.

Raff, R.A., and Kaufman, T.C. (1983). *Embryos, Genes and Evolution*. Macmillan, New York.

Raup, D.M. (1966). Geometric analysis of shell coiling: General problems. *J. Paleontol.* 40:1178–1190.

Rendel, J.M. (1967). *Canalisation and Gene Control*. Academic Press, New York/London.

Riedl, R. (1978). *Order in Livng Organisms*. Wiley, New York.

Romer, A.S. (1966). Synapsid evolution and dentition. In G. Vanderbroek (ed.), *International Colloqium on the Evolution of Lower and Nonspecialised Mammals*. Kon. Vlaamse Acad. Weten. Lett. Sch. Kunsten Belgie, Brussels.

Rosen, D.E. (1978). Darwin's demon. *Syst. Zool.* 27:370–373.

Roth, V.L. (1987). The biological basis of homology. In C.J. Humphries (ed.), *Ontogeny and Systematics*. Columbia Univ. Press, New York.

Rudwick, M.G.S. (1982). Charles Darwin in London: The integration of public and private science. *Isis* 73:186–206.

Russell, E.S. (1916). *Form and Function*. Murray, London.

Salthe, S.N. (1986). *Evolving Hierarchical Systems*. Columbia Univ. Press, New York.

Sargent, T.D., Jamrich, M., and David, I.B. (1986). Cell interactions and the control of gene activity during early development of *Xenopus laevis*. *Dev. Biol.* 114:238–246.

Sawyer, R.H., and Fallon, J.F. (eds.) (1983). *Epitheliomesenchymal Interactions in Development*. Praeger, New York.

Saxen, L., and Toivonen, S. (1962). *Primary Embryonic Induction*. Academic Press, New York.

Schaeffer, B. (1965). The role of experimentation in the origin of higher levels of organization. *Syst. Zool.* 14:318–336.

Scharf, S.R., and Gerhart, J.C. (1980). Determination of the dorso-ventral axis in eggs of *Xenopus laevis*; complete rescue of UV-impaired eggs by oblique orientation before first cleavage. *Devel. Biol.* 79:181–198.

Schmalhausen, I.I. (1949). *Factors of Evolution*. McGraw-Hill, Blakiston, New York.

Schneuwly, S., and Gehring, W.J. (1985). Homeotic transformation of thorax into head: Developmental analysis of a new antennepedia allele in *Drosophila melanogaster*. *Dev. Biol.* 108:377–386.

Schoenwolf, G.C. and Franks, M. (1984). Quantitative analyses of changes in cell shapes during bending of the avian neural plate. *Dev. Biol.* 105:257–272.

Schopf, T.J.M. (1984). Rates of evolution and the notion of "Living Fossils." *Ann. Rev. Earth Planet. Sci.* 12:245–292.

Sedgwick, A. (1894). On the law commonly known as von Baer's law; and on the significance of ancestral rudiments in embryonic development. *Q. J. Micro. Sci.* 36:38–52.

Seilacher, A. (1970). Arbeitskonzept zur Konstruktionsmorphologie. *Lethaia* 3:393–396.

Sharov, A.G. (1970). An unusual reptile from the Lower Triassic of Fergana. *Paleont. Zh* 1:112–116.

Shinomura, T., Kimata, K., Oike, Y., Maeda, N., Yano, S., and Suzuki, S. (1984). Appearance of distinct types of proteoglycan in a well-defined temporal and spatial pattern during early cartilage formation in the chick limb. *Dev. Biol.* 103:211–220.

Shubin, N.H., and Alberch, P. (1986). A morphogenetic approach to the origin and basic organization of the tetrapod limb. *Evol. Biol.* 20:319–387.

Simpson, G.G. (1953). *The Major Features of Evolution*. Columbia Univ. Press, New York.

Slack, J.M.W. (1983). *From Egg to Embryo*. Cambridge Univ. Press, London/New York.

Slavkin, Harold C., Zeichner-David, M., Snead, M.L., Graham, E.E., Samuel, N., and Ferguson, M.W.J. (1984). Amelogenesis in Reptilia: Evolutionary aspects of enamel gene products. *Symp. Zool. Soc. Lond.* 52:275–304.

Smith, J.C., and Slack, J.M.W. (1983). Dorsalization and neural induction: Properties of the organizer in *Xenopus laevis*. *J. Embryol. Exp. Morphol.* 78:299–317.

Smith, J.M. (1983). Evolution and Development. In B.C. Goodwin, N. Holder, and C.C. Wylie (eds.), *Development and Evolution*. Cambridge Univ. Press, London/New York.

Smith, J.M., Burian, R., Kauffman, S., Alberch, P., Campbell, J., Goodwin, B., Lande, R., Raup, D., and Wolpert, L. (1985). Developmental constraints and evolution. *Q. Rev. Biol.* 60:265–287.

Smith, L.J. (1985). Embryonic axis orientation in the mouse and its correlation with blastocyst relationships to the uterus. *J. Embryol. Exp. Morphol.* 89:15–35.

Sober, E. (1984). *The Nature of Selection: Evolutionary Theory in Philosophical Focus.* MIT Press, Cambridge, Mass.

Solursh, M. (1984a). Cell and matrix interactions during limb chondrogenesis *in vitro*. In R.L. Trelstad (ed.), *The Role of Extracellular Matrices in Development*. Liss, New York.

Solursh, M. (1984b). Ectoderm as a determinant of early tissue pattern in the limb bud. *Cell Diff.* 15:17–24.

Spemann, H. (1938). *Embryonic Development and Induction.* Yale Univ. Press, New Haven.

Stanley, S.M., (1975). A theory of evolution above the species level. *Proc. Natl. Acad. Sci. USA* 72:646–670.

Stark, D. and Kummer, B. (1962). Zur Ontogenes des Schimpansenschadels. *Anthropol. Anz.* 28:204–215.

Stebbins, G.L. (1974). Adaptive shifts and evolutionary: A compositionist approach. In F.J. Ayala and T. Dobzhansky (eds.), *Studies in the Philosophy of Biology*. Univ. of California Press, Berkeley.

Stein, W.D. (1980). The epigenetic address: A model for embryonic development. *J. Theor. Biol.* 82:663–677.

Stiassny, M.L.J., and Jensen, J.S. (1987). Labroid interrelationships revisited: Morphological complexity, key innovation and the study of comparative diversity. *Bull. Mus. Comp. Zool. Harvard* 151:269–319.

Stirling, R.V., and Summerbell, D. (1985). The behaviour of growing axons invading developing chick wing buds with dorsoventral or anteroposterior axis reversed. *J. Embryol. Exp. Morphol.* 85:251–269.

Sulston, J.E., Schierenberg, E., White, J., and Thomson, N. (1983). The embryonic cell lineage of the nematode *Caenorhabditis elegans*. *Dev. Biol.* 100:64–119.

Tatewaki, M., Provasoli, L.X., and Pintner, I.J. (1983). Morphogenesis of *Monostroma oxyspermum* (kutz) Doty (Chlorophyceae) in axenic culture, especially in bialgal culture. *J. Phycol.* 19:409–416.

Thiery, J.P., Dubard, J.L., and Delouvee, A. (1984). Pathways and mechanisms of avian trunk neural crest migration and localisation. *Dev. Biol.* 93:324–343.

Thompson, D'A. (1917). *On Growth and Form.* Cambridge Univ. Press, London/New York.

Thomson, K.S. (1966). The evolution of the tetrapod middle ear in the rhipidistian–tetrapod transition. *Amer. Zool.* 6:379–397.

Thomson, K.S. (1971). Adaptation and evolution of early fishes. *Q. Rev. Biol.* 46:139–166.

Thomson, K.S. (1976). Explanation of large-scale extinctions of lower vertebrates. *Nature* 261:578–580.

Thomson, K.S. (1977). The pattern of diversification among fishes. In A. Hallam (ed.), *Patterns of Evolution, as Illustrated in the Fossil Record*. Elseview–North Holland, Amsterdam/New York.

Thomson, K.S. (1982). The meanings of evolution. *Am. Sci.* 70:529–531.

Thomson, K.S. (1983). Sine scientia ars nihil est? *Am Sci.* 71:247–299.

Thomson, K.S. (1984). Reductionism and other -isms in biology. *Am. Sci.* 72:388–390.

Thomson, K.S. (1986). On the relationship between development and evolution. *Oxford Surv. Evol. Biol.* 2:319–333.

Thomson, K.S. (1987a). Living fossils. *Paleobiology* 12:495–498.

Thomson, K.S. (1987b). Speculations concerning the role of the neural crest in the morphogenesis and evolution of the vertebrate skeleton. In P.F.A. Maderson (ed.), *The Neural Crest.* Wiley, New York.

Thomson, K.S. (1988). Head segmentation: Homology, morphogenesis and evolution. In T. Horder, and R. Presley (eds.), *Vertebrate Head Segmentation.* Oxford Univ. Press, London/ New York (in press).

Thorogood, P. (1983). Morphogenesis of Cartilage. In *Cartilage,* Vol. 2, *Development, Differentiation, and Growth.* Academic Press, New York/London.

Thorogood, P., Bee, J., and v.d. Mark, K. (1986). Transient expression of collagen Type II at epitheliomesenchymal interfaces during morphogenesis of the cartilagenous neurocranium. *Dev. Biol.* 116:497–509.

Tiedman, H. (1967). Biochemical aspects of primary induction and determination. In R. Weber (ed.), *The Biochemistry of Animal Development.* Academic Press, New York/London.

Toivonen, S. (1978). Regionalisation of the embryo. In O. Nakamura, and S. Toivonen (eds.), *The Organizer.* Elsevier–North Holland, Amsterdam/New York.

Toivonen, S. (1979). Transmission problems in primary induction. *Differentiation* 15:177–181.

Tosney, K.W., Watanabe, M., Landmesser, L., and Rutishauser, R. (1986). The distribution of NCAM in the chick hindlimb during axon outgrowth and synaptogenesis. *Dev. Biol.* 114:437–452.

Trelstadt, R.L. (ed.) (1984). *The Role of Extracellular Matrix in Development.* Alan R. Liss, New York.

Tuckett, F., and Morriss-Kay, G.M. (1985). The kinetic behaviour of the cranial neural epithelium during neurulation in the rat. *J. Embryol. Exp. Morphol.* 85:111–119.

Turing, A. (1952). The chemical basis of morphogenesis. *Philos. Trans. R. Soc. Lond. [Biol.]* 237:37–72.

Ubbels, G.A. (1977). Symmetrisation of the fertilised eggs of *Xenopus laevis* studied by cytological, cytochemical and ultrastructural methods. *Mem. Soc. Zool. Fr.* 41:103–116.

Ubbels, G.A., Hara, K., Koster, C.H., and Kirschner, M.W. (1983). Evidence for a functional role of the cytoskeleton in determination of the dorsoventral axis in *Xenopus laevis J. Embryol. Exp. Morphol.* 77:15–37.

van Dam, W.I., and Verdonk, N.H. (1982). The morphogenetic significance of the first quartet micromeres for the development of the snail *Bithyria tentaculata. Wm. Roux' Arch. Dev. Biol.* 191:112–118.

van Valen, L.M. (1973). A new evolutionary law. *Evol. Theory* 1:1–30.

van Valen, L.M. (1974). A natural model for the origin of some higher taxa. *J. Herpetol.* 8:109–121.

van Valen, L.M. (1976). Energy and evolution. *Evol. Theory* 1:179–229.

van Valen, L.M. (1985). A theory of origination and extinction. *Evol. Theory* 7:133–142.

Vincent, J., Oster, G.F., and Gerhart, J.C. (1986). Kinematics of gray crescent formation in *Xenopus* eggs: The displacement of subcortical cytoplasm relative to the egg surface. *Dev. Biol.* 113:484–500.

von Baer, K.E. (1828). *Entwicklungsgeschichte der Tiere: Beobachtung und Reflexion.* Borntraeger, Konigsberg.

Vrba, E.S., and Eldredge, N. (1984). Individuals, hierarchies and processes: Towards a more complete evolutionary theory. *Paleobiology* 10:146–171.

Waddington, C.H. (1961). Genetic assimilation. *Adv. Genet.* 10:257–290.

Waddington, C.H. (1967). *The Strategy of the Genes.* Allen & Unwin, London.

Waddington, C.H. (1975). *The Evolution of an Evolutionist.* Cornell Univ. Press, Ithaca, N.Y.

Wake, D.B. (1966). Comparative osteology and evolution of the lungless salamanders, family Plethodontidae. *Mem. South. Calif. Acad. Sci.* 4:1–111.

Wake, D.B. (1982a). Functional and developmental constraints and opportunities in the evolution of feeding systems in *Urodeles.* In D. Mossakowski and G. Roth (eds.), *Environmental Adaptation and Evolution.* Sustrav Fischer, Stuttgart, New York.

Wake, D.B. (1982b). Functional and evolutionary morphology. *Persp. in Biol. and Med.* 25:603–620.

Wake, M.H., Bemis, K., and Schwenk, K. (1983). Morphology and function of the feeding apparatus in *Dermophis mexicanus* (Amphibia: Gymnophiona). *Zool. J. Linn. Soc. Lond.* 77:75–96.

Walpert, J. (1969). Positional information and the spatial pattern of cellular differentiation. *J. Theor. Biol.* 25:1–47.

Webster, G.C., and Goodwin, B.C. (1982). The origin of species: A structuralist approach. *J. Soc. Biol. Struct.* 5:15–47.

Wessells, W.K. (1982). A catalogue of processes responsible for metazoan morphogenesis. In J.T. Bonner (ed.), *Evolution and Development.* Springer-Verlag, Berlin/New York.

West, J.D., and Green, J.F. (1983). The transition from oocyte-coded to embryo-coded glucose phosphate isomerase in the early mouse embryo. *J. Embryol. Exp. Morphol.* 78:127–140.

Whittacker, J.R. (1980). Acetylcholinesterase development in extra cells caused by changing the distribution of myoplasm in ascidian embryos. *J. Embryol. Exp. Morphol.* 55:343–354.

Williams, G.C. (1986). A defense of reductionism in evolutionary biology. *Oxford Surv. Evol. Biol.* 2:1–27.

Wolff, E. (1958). Le principe de competition. *Bull. Soc. Zool. Fr.* 83:13–25.

Wolpert, L. (1969). Positional information and the spatial pattern of cellular information. *J. Theor. Biol.* 25:1–47.

Wolpert, L. (1981). Positional information and pattern formation. *Philos. Trans. R. Soc. Lond.* [*Biol.*] 295:441–450.

Wolpert, L. (1983). Pattern formation and change. In J.T. Bonner (ed.), *Evolution and Development.* Springer-Verlag, Berlin/New York.

Zanetti, N.C., and Solursh, M. (1986). Epithelial effects on limb chondrogenesis involve extracellular matrix and cell shape. *Dev. Biol.* 113:110–118.

Zust, B., and Dixon, K.E. (1977). Events in the germ cell lineage after entry of the primordial germ cells into the gentical ridges in normal and irradiated *Xenopus laevis. J. Embryol. Exp. Morphol.* 41:33–46.

Index